# Forensic Psychology

# Forensic Psychology

Neera Roy

**AMIGA PRESS. INC**
**Delhi -110051**

**AMIGA PRESS. INC**
H.O :- 28-A, Indra Park, Chander Nagar
Delhi :- 110051 (India)
Tel :-   011-22504028, +91-98736-57000
Email :-  amigapress@gmail.com
Website:  www.amigapress.co.in

This edition is published by Amiga Press.Inc

ISBN : 978-93-84533-08-3

Copyright© Reserved
First Published 2018

Publisher's note:
This publication may not reproduced, stored in a retrieval system or transmitted in any form or by any means, photocopying , mechanical, electronic, recording or otherwise, without the prior permission from publication house.

Printed and bound in India by Replika Press Pvt. Ltd.

# Preface

Forensic psychology is the intersection between psychology and the justice system. It involves understanding fundamental legal principles, particularly with regard to expert witness testimony and the specific content area of concern (e.g., competence to stand trial, child custody and visitation, or workplace discrimination), as well as relevant jurisdictional considerations (e.g., in the United States, the definition of insanity in criminal trials differs from state to state) in order to be able to interact appropriately with judges, attorneys and other legal professionals.

Forensic psychologists are different from regular therapists or counsellors because they are not as interested in the client-therapist relationship or building rapport. Instead, forensic psychologists attempt to evaluate individuals according to guiding laws and statutes. They are adept at recognizing deception. Forensic psychologists also profile criminals and explain the thought and behaviour patterns of particular crimes and groups of people (e.g. serial killers, pedophiles, arsonists and terrorists).

Forensic psychology is the intersection between psychology and the courtroom—criminal, civil, family and Federal.]. It involves understanding and applying accepted principles and practices in psychology with relevant legal issues jurisdictions in order to be able to interact appropriately with judges, attorneys and other legal professionals.

Forensic science is the application of science to criminal and civil laws. Forensic scientists collect, preserve, and analyse scientific evidenceduring the course of an investigation. While some forensic scientists travel to the scene to collect the evidence themselves, others occupy a purely laboratory role, performing analysis on objects brought to them by other individuals. In addition to their laboratory role, forensic scientists testify as expert witnesses in both criminal and civil cases and can work for either the prosecution or the defense. While any field could technically be forensic, certain sections have developed over time to encompass the majority of forensically related cases.

Digital forensics investigations have a variety of applications. The most common is to support or refute a hypothesis before criminal or civil (as part of the electronic discovery process) courts. Forensics may also feature in the private sector; such as during internal corporate investigations or intrusion investigation (a specialist probe into the nature and extent of an unauthorized network intrusion). The technical aspect of an investigation is divided into several sub-branches, relating to the type of digital devices involved; computer forensics, network forensics, forensic data analysis and mobile device forensics. The typical forensic process encompasses the seizure, forensic imaging (acquisition) and analysis of digital media and the production of a report into collected evidence.

An overriding issue in any type of forensic assessment is the issue of malingering and deception. A defendant may be intentionally faking a mental illness or may be exaggerating the degree of symptomatology. The forensic psychologist must always keep this possibility in mind. It is important if malingering is suspected to observe the defendant in other settings as it is difficult to maintain false symptoms consistently over time. In some cases, the court views malingering or feigning illness as obstruction of justice and sentences the defendant accordingly. In United States v. Binion, malingering or feigning illness during a competency evaluation was held to be obstruction of justiceand led to an enhanced sentence. As such, fabricating mental illness in a competency-to-stand-trial assessment now can be raised to enhance the sentencing level following a guilty plea.

The book gives budding writers, and anyone else with an interest in this subject, a solid grounding in the history, terminology, and techniques of forensic psychology.

—*Editor*

# Contents

| | | |
|---|---|---|
| | *Preface* | (v) |
| 1. | **Introduction** | 1 |
| | Distinction between Forensic and Therapeutic Evaluation; Litigation Science; Early Research in Forensic Psychology; Forensic Psychology in the Courts; Forensic Psychology Practice | |
| 2. | **Digital Forensics** | 13 |
| | Computer Forensics; Network Forensics; Digital Forensic Process; Forensic process ; Mobile Device Forensics | |
| 3. | **Forensic Sciences** | 33 |
| | History ; Mitigating Circumstances; Forensic Evidence Case Vignette | |
| 4. | **Criminal Psychology and "Fear of Crime"** | 53 |
| | Crime Assessment Stage ; Criminal Profile Stage ; Criminal Psychologist; Crime and Criminology; Computational Criminology; Biological Theories in Criminology; Causes of Crime; Heredity and Brain Activity; Crimes against Women; Consequences of Crimes; Discouraging the Choice of Crime; Criminal Profiling from Crime Scene Analysis | |
| 5. | **Psychological Testing** | 101 |
| | Psychometric and Ethical Standards; Psychometric Testing; Post-Assessment Phase; Psychological Tests | |
| 6. | **Cognitive Assessment** | 126 |
| | Structured Interview of Reported Symptoms (SIRS) ; Competency Assessment; Violence Risk Assessment; Psychological Tests | |
| 7. | **Psychology of Criminal Behaviour** | 137 |
| | Definition and Measurement of Criminal Behaviour; Criminal Behaviour and Personality Disorders; Personality Disorders and | |

Traits; Psychological and Osychiatric Theories of Criminal Behaviour

8. **Criminal Mind and Offender Profiling**   160
   Criminal Minds; Automated Fingerprint Identification System; Psychological Theories of Crime; Criminal Profiling ; Psychological Profiling ; Criminal Profile ; Offender Typologies

9. **Crime and Forensic Science**   182
   Investigation of Crime ; The Control of Crime and Deviance ; General Theory of Crime; Learning Theories of Crime; Class and Crime ; Shifts In Focus

   *Bibliography*   209

   *Index*   211

# 1

# Introduction

Forensic psychology is an application of psychology to legal issues and the criminal justice system. In order to understand what forensic psychology encompasses, consider the types of questions forensic psychologists answer in regard to the law and court system.
- Is a particular individual competent to stand trial?
- What was the individual's state of mind when he or she committed the act? (e.g. sane versus insane, accidental versus pre-meditated)
- Was the individual solely responsible for his or her actions or were they being manipulated or threatened to commit the act?
- Are the witnesses or expert witnesses credible?
- Is the jury objective or impartial?
- Are the lawyers acting in their client's best interest?
- Is anyone involved malingering (that's a fancy word for lying)?
- Is the sentence/punishment appropriate for the crime and or state of mind of the defendant?
- Can witness testimony be trusted as fact? Do we accept "recovered memory" as evidence of a crime?
- In situations where there are several culprits, how we determine who is the most to blame and therefore should get the heaviest sentence or should they all be treated the same?
- Should children that commit crimes be charged similarly to adults?

Forensic psychologists are different from regular therapists or counsellors because they are not as interested in the client-therapist relationship or building rapport. Instead, forensic psychologists attempt to evaluate individuals according to guiding laws and statutes. They are adept at recognizing deception. Forensic psychologists also profile criminals and explain the thought and behaviour patterns of particular crimes and groups of people (e.g. serial killers, pedophiles, arsonists and terrorists).

Forensic psychology is the intersection between psychology and the courtroom—criminal, civil, family and Federal.]. It involves understanding and applying accepted principles and practices in psychology with relevant legal issues jurisdictions in order to be able to interact appropriately with judges, attorneys and other legal professionals.

An important aspect of forensic psychology is the ability to testify in court, reformulating psychological findings into the legal language of the courtroom, providing information to legal personnel in a way that can be understood.

Further, in order to be a credible witness, for example in the United States, the forensic psychologist must understand the philosophy, rules and standards of the American judicial system. Primary is an understanding of the adversarial system. There are also rules about hearsay evidence and most importantly, the exclusionary rule. Lack of a firm grasp of these procedures will result in the forensic psychologist losing credibility in the courtroom. A forensic psychologist can be trained in clinical, social, organizational or any other branch of psychology. In the United States, the salient issue is the designation by the court as an expert witness by training, experience or both by the judge. Generally, a forensic psychologist is designated as an expert in a particular jurisdiction. The number of jurisdictions in which a forensic psychologist qualifies as an expert increases with experience and reputation. Forensic neuropsychologists are generally asked to appear as expert witnesses in court to discuss cases that involves issues with the brain or brain damage. They also deal with issues of whether a person is legally competent enough to stand trial.

According to R.J. Gregory in *Psychological Testing: History, Principles, and Application*, the main roles of a psychologist in the court system are eight-fold:
- Evaluation of possible malingering
- Assessment of mental state for insanity plea
- Competency to stand trial
- Prediction of violence and assessment of risk
- Evaluation of child custody in divorce
- Assessment of personal injury
- Interpretation of polygraph data
- Specialized forensic personality assessment

Questions asked by the court of a forensic psychologist are generally not questions regarding psychology but are legal questions and the response must be in language the court understands. For example, a forensic psychologist is frequently appointed by the court to assess a defendant's competency to stand trial. The court also frequently appoints a forensic psychologist to assess

# Introduction

the state of mind of the defendant at the time of the offence. This is referred to as an evaluation of the defendant's sanity or insanity (which relates to criminal responsibility) at the time of the offence. These are not primarily psychological questions but rather legal ones. Thus, a forensic psychologist must be able to translate psychological information into a legal framework.

Forensic psychologists provide sentencing recommendations, treatment recommendations, and any other information the judge requests, such as information regarding mitigating factors, assessment of future risk, and evaluation of witness credibility.

Forensic psychology also involves training and evaluating police or other law enforcement personnel, providing law enforcement with criminal profiles and in other ways working with police departments.

Forensic psychologists work both with the Public Defender, the States Attorney, and private attorneys. Forensic psychologists may also help with jury selection.

## DISTINCTION BETWEEN FORENSIC AND THERAPEUTIC EVALUATION

A forensic psychologist's interactions with and ethical responsibilities to the client differ widely from those of a psychologist dealing with a client in a clinical setting.

- Scope. Rather than the broad set of issues a psychologist addresses in a clinical setting, a forensic psychologist addresses a narrowly defined set of events or interactions of a nonclinical nature.
- Importance of client's perspective. A clinician places primary importance on understanding the client's unique point of view, while the forensic psychologist is interested in accuracy, and the client's viewpoint is secondary.
- Voluntariness. Usually in a clinical setting a psychologist is dealing with a voluntary client. A forensic psychologist evaluates clients by order of a judge or at the behest of an attorney.
- Autonomy. Voluntary clients have more latitude and autonomy regarding the assessment's objectives. Any assessment usually takes their concerns into account. The objectives of a forensic examination are confined by the applicable statutes or common law elements that pertain to the legal issue in question.
- Threats to validity. While the client and therapist are working toward a common goal, although unconscious distortion may occur, in the forensic context there is a substantially greater likelihood of intentional and conscious distortion.

- Relationship and dynamics. Therapeutic interactions work toward developing a trusting, empathic therapeutic alliance, a forensic psychologist may not ethically nurture the client or act in a "helping" role, as the forensic evaluator has divided loyalties and there are substantial limits on confidentiality he can guarantee the client. A forensic evaluator must always be aware of manipulation in the adversary context of a legal setting. These concerns mandate an emotional distance that is unlike a therapeutic interaction.
- Pace and setting. Unlike therapeutic interactions which may be guided by many factors, the forensic setting with its court schedules, limited resources, and other external factors, place great time constraints on the evaluation without opportunities for reevaluation. The forensic examiner focuses on the importance of accuracy and the finality of legal dispositions.

## LITIGATION SCIENCE

Litigation science describes analysis or data developed or produced *expressly* for use in a trial, versus those produced in the course of independent research. This distinction was made by the U.S. 9th Circuit Court of Appeals when evaluating the admissibility of experts. This uses demonstrative evidence, which is evidence created in preparation of trial by attorneys or paralegals.

### Examples in Popular Culture

Sherlock Holmes, the fictional character created by Sir Arthur Conan Doyle in works produced from 1887 to 1915, used forensic science as one of his investigating methods.

Conan Doyle credited the inspiration for Holmes on his teacher at the medical school of the University of Edinburgh, the gifted surgeon and forensic detective Joseph Bell. Decades later, the comic strip *Dick Tracy* also featured a detective using a considerable number of forensic methods, although sometimes the methods were more fanciful than actually possible.

Defense attorney Perry Mason occasionally used forensic techniques, both in the novels and television series. Popular television series focusing on crime detection, including *The Mentalist, CSI, Cold Case, Bones, Law & Order, NCIS, Criminal Minds, Silent Witness, Dexter,* and *Waking the Dead*, depict glamorized versions of the activities of 21st century forensic scientists. Some claim these TV shows have changed individuals' expectations of forensic science, an influence termed the "CSI effect". Non-fiction TV shows such as Forensic Files, The New Detectives, American Justice, and Dayle Hinman's Body of Evidence have also popularized the area of forensic science.

## Controversies

Questions about forensic science, fingerprint evidence and the assumption behind these disciplines have been brought to light in some publications, the latest being an article in the *New York Post*. The article stated that "No one has proved even the basic assumption: That everyone's fingerprint is unique." The article also stated that "Now such assumptions are being questioned — and with it may come a radical change in how forensic science is used by police departments and prosecutors."

On June 25, 2009 the Supreme Court issued a 5-to-4 decision in Melendez-Diaz v. Massachusetts stating that crime laboratory reports may not be used against criminal defendants at trial unless the analysts responsible for creating them give testimony and subject themselves to cross-examination.

The Supreme Court cited the National Academies report *Strengthening Forensic Science in the United States* in their decision. Writing for the majority, Justice Antonin Scalia referenced the National Research Council report in his assertion that "Forensic evidence is not uniquely immune from the risk of manipulation."

## EARLY RESEARCH IN FORENSIC PSYCHOLOGY

J. McKeen Cattell conducted some of the earliest research on the psychology of testimony. He posed a series of questions to students at Columbia University, asking them to provide a response and rate their degree of confidence in their answer (1895). Cattell's results indicated a surprising degree of inaccuracy, which generated interest among other psychologists who went on to conduct experiments on eyewitness testimony.

Inspired by Cattell's work, Alfred Binet replicated Cattell's research and studied the results of other psychology experiments that applied to law and criminal justice. His work in intelligence testing was also important to the development of forensic psychology, as many future assessment tools were based on his work.

Psychologist William Stern also studied witness recall. In one experiment, students were asked to summarize a dispute they witnessed between two classmates. Stern discovered that errors were common among the witnesses, concluding that emotions decrease the accuracy of witness recall. Stern continued to study issues surrounding testimony and later established the first academic journal devoted to applied psychology.

## FORENSIC PSYCHOLOGY IN THE COURTS

During this time, psychologists were also beginning to act as expert witnesses in criminal trials throughout Europe. In 1896, a psychologist by the name of Albert von Schrenck-Notzing testified at a murder trial about the effects of suggestibility on witness testimony. Hugo Munsterberg's ardent belief that psychology had practical applications in everyday life also contributed to the development of forensic psychology. In 1908, Munsterberg published his book *On the Witness Stand*, advocating the use of psychology in legal matters. Despite his contributions, Munsterberg was generally disliked by many of his peers in psychology and by much of the legal community.

Stanford psychologist Lewis Terman began applying psychology to law enforcement in 1916. After revising Binet's intelligence test, the new Stanford-Binet test was used to assess the intelligence of job candidates for law enforcement positions. In 1917, psychologist William Marston (a student of Munsterberg) found that systolic blood pressure had a strong correlation to lying. This discovery would later lead to the design of the modern polygraph detector.

Marston testified in 1923 in the case of *Frye vs. United States*. This case is significant because it established the precedent for the use of expert witnesses in courts. The Federal Court of Appeals determined that a procedure, technique, or assessment must be generally accepted within its field in order to be used as evidence.

## Post-WWII Growth

Significant growth in American forensic psychology did not happen until after World War II. Psychologist served as expert witnesses, but only in trials that weren't perceived as infringing on medical specialists, who were seen as more credible witnesses. In the 1940 case of the *People vs. Hawthorne*, the courts ruled that the standard for expert witnesses was in the extent of knowledge of a subject, not in whether or not the witness had a medical degree.

In the landmark 1954 case of Brown vs. Board of Education, several psychologists testified for both the plaintiffs and the defendants. Later, the courts gave support to psychologists serving as mental illness experts in the case of *Jenkins vs. United States* (1962). Forensic psychology has continued to grow and evolve during the past three decades. Increasing numbers of graduate programs offer dual degrees in psychology and law, while other offer specialized degrees emphasized forensic psychology. In 2001, the American Psychological Association officially recognized forensic psychology as a specialization within psychology.

## Forensic Science

Forensic science (often shortened to forensics) is the application of a broad spectrum of sciences to answer questions of interest to a legal system. This may be in relation to a crime or a civil action. Besides its relevance to a legal system, more generally *forensics* encompasses the accepted scholarly or scientific methodology and norms under which the facts regarding an event, or an artifact, or some other physical item (such as a corpse) are ascertained as being the case. In that regard the concept is related to the notion of authentication, where by an interest outside of a legal form exists in determining whether an object is what it purports to be, or is alleged as being.

The word *forensic* comes from the Latin adjective *forensis*, meaning "of or before the forum". In Roman times, a criminal charge meant presenting the case before a group of public individuals in the forum. Both the person accused of the crime and the accuser would give speeches based on their side of the story. The individual with the best argument and delivery would determine the outcome of the case. This origin is the source of the two modern usages of the word *forensic* – as a form of legal evidence and as a category of public presentation. In modern use, the term "forensics" in place of "forensic science" can be considered incorrect as the term "forensic" is effectively a synonym for "legal" or "related to courts". However, the term is now so closely associated with the scientific field that many dictionaries include the meaning that equates the word "forensics" with "forensic science".

## Modern History

In sixteenth century Europe, medical practitioners in army and university settings began to gather information on cause and manner of death. Ambroise Pare, a French army surgeon, systematically studied the effects of violent death on internal organs. Two Italian surgeons, Fortunato Fidelis and Paolo Zacchia, laid the foundation of modern pathology by studying changes which occurred in the structure of the body as the result of disease. In the late 1700s, writings on these topics began to appear. These included: *A Treatise on Forensic Medicine and Public Health* by the French physician Fodere, and *The Complete System of Police Medicine* by the German medical expert Johann Peter Franck.

In 1776, Swedish chemist Carl Wilhelm Scheele devised a way of detecting arsenous oxide, simple arsenic, in corpses, although only in large quantities. This investigation was expanded, in 1806, by German chemist Valentin Ross, who learned to detect the poison in the walls of a victim's stomach, and by English chemist James Marsh, who used chemical processes to confirm arsenic as the cause of death in an 1836 murder trial.

Two early examples of English forensic science in individual legal proceedings demonstrate the increasing use of logic and procedure in criminal investigations. In 1784, in Lancaster, England, John Toms was tried and convicted for murdering Edward Culshaw with a pistol. When the dead body of Culshaw was examined, a pistol wad (crushed paper used to secure powder and balls in the muzzle) found in his head wound matched perfectly with a torn newspaper found in Toms' pocket. In Warwick, England, in 1816, a farm labourer was tried and convicted of the murder of a young maidservant. She had been drowned in a shallow pool and bore the marks of violent assault. The police found footprints and an impression from corduroy cloth with a sewn patch in the damp earth near the pool. There were also scattered grains of wheat and chaff. The breeches of a farm labourer who had been threshing wheat nearby were examined and corresponded exactly to the impression in the earth near the pool. Later in the 20th century, several British pathologists, Bernard Spilsbury, Francis Camps, Sydney Smith and Keith Simpson would pioneer new forensic science methods in Britain. In 1909 Rodolphe Archibald Reiss founded the first school of forensic science in the world: the "Institut de police scientifique" in the University of Lausanne (UNIL).

Subdivisions:
- Computational forensics concerns the development of algorithms and software to assist forensic examination.
- Criminalistics is the application of various sciences to answer questions relating to examination and comparison of biological evidence, trace evidence, impression evidence (such as fingerprints, footwear impressions, and tire tracks), controlled substances, ballistics, firearm and toolmark examination, and other evidence in criminal investigations. Typically, evidence is processed in a crime lab.
- Digital forensics is the application of proven scientific methods and techniques in order to recover data from electronic/digital media. Digital Forensic specialists work in the field as well as in the lab.
- Forensic anthropology is the application of physical anthropology in a legal setting, usually for the recovery and identification of skeletonized human remains.
- Forensic archaeology is the application of a combination of archaeological techniques and forensic science, typically in law enforcement.
- Forensic DNA analysis takes advantage of the uniqueness of an individual's DNA to answer forensic questions such as paternity/maternity testing or placing a suspect at a crime scene, e.g. in a rape investigation.
- Forensic entomology deals with the examination of insects in, on, and around human remains to assist in determination of time or location of death. It is also possible to determine if the body was moved after death.
- Forensic geology deals with trace evidence in the form of soils, minerals and petroleums.

## Introduction

- Forensic meteorology is a site specific analysis of past weather conditions for a point of loss.
- Forensic odontology is the study of the uniqueness of dentition better known as the study of teeth.
- Forensic pathology is a field in which the principles of medicine and pathology are applied to determine a cause of death or injury in the context of a legal inquiry.
- Forensic psychology is the study of the mind of an individual, using forensic methods. Usually it determines the circumstances behind a criminal's behaviour.
- Forensic seismology is the study of techniques to distinguish the seismic signals generated by underground nuclear explosions from those generated by earthquakes.
- Forensic toxicology is the study of the effect of drugs and poisons on/in the human body.
- Forensic document examination or questioned document examination answers questions about a disputed document using a variety of scientific processes and methods. Many examinations involve a comparison of the questioned document, or components of the document, to a set of known standards. The most common type of examination involves handwriting wherein the examiner tries to address concerns about potential authorship.
- Forensic video analysis is the scientific examination, comparison, and evaluation of video in legal matters.
- Forensic engineering is the scientific examination and analysis of structures and products relating to their failure or cause of damage.
- Forensic limnology is the analysis of evidence collected from crime scenes in or around fresh water sources. Examination of biological organisms, particularly diatoms, can be useful in connecting suspects with victims.
- Forensic Accounting is the study and interpretation of accounting evidence
- Forensic Botany is the study of plant life in order to gain information regarding possible crimes
- Forensic Dactyloscopyis the study of fingerprints (patent and latent fingerprints)
- Trace Evidence Analysis is the analysis and comparison of trace evidence including (glass, paint, fibers, hair), etc.
- Forensic Chemistry is the study of detection and identification of illicit drugs, accelerants used in arson cases, explosive and gunshot resides

- Forensic Optometry is the study of glasses and other eye wear relating to crime scenes and criminal investigations
- Forensic Serology is the study of blood groups and blood for identification purposes following a crime, the study of how blood splaters, and the analysis of blood stains
- Forensic Linguistics deals with anything in the legal system that requires linguistic expertise.

## Questionable Techniques

Some forensic techniques, believed to be scientifically sound at the time they were used, have turned out later to have much less scientific merit, or none. Some such techniques include:
- Comparative bullet-lead analysis was used by the FBI for over four decades, starting with the John F. Kennedy assassination in 1963. The theory was that each batch of ammunition possessed a chemical makeup so distinct that a bullet could be traced back to a particular batch, or even a specific box. However, internal studies and an outside study by the National Academy of Sciences found that the technique was unreliable, and the FBI abandoned the test in 2005.
- Forensic dentistry has come under fire; in at least two cases, bite mark evidence has been used to convict people of murder who were later freed by DNA evidence. A 1999 study by a member of the American Board of Forensic Odontology found a 63 percent rate of false identifications and is commonly referenced within online news stories and conspiracy websites. However, the study was based on an informal workshop during an ABFO meeting which many members did not consider a valid scientific setting.

## FORENSIC PSYCHOLOGY PRACTICE

The forensic psychologist views the client or defendant from a different point of view than does a traditional clinical psychologist. Seeing the situation from the client's point of view or "empathizing" is not the forensic psychologist's task. Traditional psychological tests and interview procedure are not sufficient when applied to the forensic situation. In forensic evaluations, it is important to assess the consistency of factual information across multiple sources.

Forensic evaluators must be able to provide the source on which any information is based. While psychologists infrequently have to be concerned about malingering or feigning illness in a non-criminal clinical setting, a forensic psychologist must be able to recognize exaggerated or faked symptoms. Malingering exists on a continuum so the forensic psychologist must be skilled in recognizing varying degrees of feigned symptoms.

# Introduction

Forensic psychologists perform a wide range of tasks within the criminal justice system. By far the largest is that of preparing for and providing testimony in the court room. This task has become increasingly difficult as attorneys have become sophisticated at undermining psychological testimony. Evaluating the client, preparing for testimony, and the testimony itself require the forensic psychologist to have a firm grasp of the law and the legal situation at issue in the courtroom, using the Crime Classification Manual and other sources. This knowledge must be integrated with the psychological information obtained from testing, psychological and mental status exams, and appropriate assessment of background materials, such as police reports, prior psychiatric or psychological evaluations, medical records and other available pertinent information.

## Malingering

An overriding issue in any type of forensic assessment is the issue of malingering and deception. A defendant may be intentionally faking a mental illness or may be exaggerating the degree of symptomatology. The forensic psychologist must always keep this possibility in mind. It is important if malingering is suspected to observe the defendant in other settings as it is difficult to maintain false symptoms consistently over time. In some cases, the court views malingering or feigning illness as obstruction of justice and sentences the defendant accordingly. In United States v. Binion, malingering or feigning illness during a competency evaluation was held to be obstruction of justiceand led to an enhanced sentence. As such, fabricating mental illness in a competency-to-stand-trial assessment now can be raised to enhance the sentencing level following a guilty plea.

## Competency Evaluations

If there is a question of the accused's competency to stand trial, a forensic psychologist is appointed by the court to examine and assess the individual. The individual may be in custody or may have been released on bail. Based on the forensic assessment, a recommendation is made to the court whether or not the defendant is competent to proceed to trial.

If the defendant is considered incompetent to proceed, the report or testimony will include recommendations for the interim period during which an attempt at restoring the individual's competency to understand the court and legal proceedings, as well as participate appropriately in their defence will be made.

Often, this is an issue of committed, on the advice of a forensic psychologist, to a psychiatric treatment facility until such time as the individual is deemed competent.

As a result of Ford v. Wainwright, a case by a Florida inmate on death row that was brought before the Supreme Court of the United States, forensic psychologists are appointed to assess the competency of an inmate to be executed in death penalty cases.

## Sanity Evaluations

The forensic psychologist may also be appointed by the court to evaluate the defendant's state of mind at the time of the offence. These are defendants who the judge, prosecutor or public defender believe, through personal interaction with the defendant or through reading the police report, may have been significantly impaired at the time of the offence. In other situations, the defence attorney may decide to have the defendant plead not guilty by reason of insanity. In this case, usually the court appoints forensic evaluators and the defence may hire their own forensic expert.

In actual practice, this is rarely a plea in a trial. A plea for insanity is actually used in only 1 in 1000 cases. Assessments that would be used can include the Mental State at time of Offence (MSO), an assessment that judges the individual's mental state when the offence was committed, helping to decide whether they should be held liable for the crime. The individual can also plea 'Not Guilty by Reason of Insanity' (NGRI) or 'Guilty but Mentally Ill' (GBMI), cases where the individual will start their sentence in a mental health facility and then complete it in correctional facility. Usually any judgments about the defendant's state of mind at the time of the offence are made by the court before the trial process begins.

## Sentence Mitigation

Even in situations where the defendant's mental disorder does not meet the criteria for a not guilty by reason of insanity defence, the defendant's state of mind at the time, as well as relevant past history of mental disorder and psychological abuse can be used to attempt a mitigation of sentence. The forensic psychologist's evaluation and report is an important element in presenting evidence for sentence mitigation. In *Hamblin v. Mitchell*, 335 F.3d 482 (6th Cir. 2003), the Sixth Circuit Court of Appeals reversed the decision of a lower court because counsel did not thoroughly investigate the defendant's mental history in preparation for the sentencing phase of the trial.

# 2

# Digital Forensics

Digital forensics (sometimes known as digital forensic science) is a branch of forensic science encompassing the recovery and investigation of material found in digital devices, often in relation to computer crime. The term digital forensics was originally used as a synonym for computer forensics but has expanded to cover investigation of all devices capable of storing digital data. With roots in thepersonal computing revolution of the late 1970s and early 1980s, the discipline evolved in a haphazard manner during the 1990s, and it was not until the early 21st century that national policies emerged.

Digital forensics investigations have a variety of applications. The most common is to support or refute a hypothesis before criminal or civil (as part of the electronic discovery process) courts. Forensics may also feature in the private sector; such as during internal corporate investigations or intrusion investigation (a specialist probe into the nature and extent of an unauthorized network intrusion).

The technical aspect of an investigation is divided into several sub-branches, relating to the type of digital devices involved; computer forensics, network forensics, forensic data analysis and mobile device forensics. The typical forensic process encompasses the seizure, forensic imaging (acquisition) and analysis of digital media and the production of a report into collected evidence.

As well as identifying direct evidence of a crime, digital forensics can be used to attribute evidence to specific suspects, confirm alibis or statements, determine intent, identify sources (for example, in copyright cases), or authenticate documents. Investigations are much broader in scope than other areas of forensic analysis (where the usual aim is to provide answers to a series of simpler questions) often involving complex time-lines or hypotheses.

## History

Prior to the 1980s crimes involving computers were dealt with using existing laws. The first computer crimes were recognized in the 1978 Florida

Computer Crimes Act, which included legislation against the unauthorized modification or deletion of data on a computer system. Over the next few years the range of computer crimes being committed increased, and laws were passed to deal with issues of copyright, privacy/harassment (e.g., cyber bullying, cyber stalking, and online predators) and child pornography. It was not until the 1980s that federal laws began to incorporate computer offences. Canada was the first country to pass legislation in 1983. This was followed by the US Federal *Computer Fraud and Abuse Act* in 1986, Australian amendments to their crimes acts in 1989 and the British *Computer Abuse Act* in 1990.

## *1980s–1990s: Growth of the field*

The growth in computer crime during the 1980s and 1990s caused law enforcement agencies to begin establishing specialized groups, usually at the national level, to handle the technical aspects of investigations. For example, in 1984 the FBI launched a *Computer Analysis and Response Team* and the following year a computer crime department was set up within the British Metropolitan Police fraud squad. As well as being law enforcement professionals, many of the early members of these groups were also computer hobbyists and became responsible for the field's initial research and direction.

One of the first practical (or at least publicized) examples of digital forensics was Cliff Stoll's pursuit of hacker Markus Hess in 1986. Stoll, whose investigation made use of computer and network forensic techniques, was not a specialized examiner. Many of the earliest forensic examinations followed the same profile.

Throughout the 1990s there was high demand for these new, and basic, investigative resources. The strain on central units lead to the creation of regional, and even local, level groups to help handle the load. For example, the British National Hi-Tech Crime Unit was set up in 2001 to provide a national infrastructure for computer crime; with personnel located both centrally in London and with the variousregional police forces (the unit was folded into the Serious Organised Crime Agency (SOCA) in 2006).

During this period the science of digital forensics grew from the ad-hoc tools and techniques developed by these hobbyist practitioners. This is in contrast to other forensics disciplines which developed from work by the scientific community. It was not until 1992 that the term "computer forensics" was used in academic literature (although prior to this it had been in informal use); a paper by Collier and Spaul attempted to justify this new discipline to the forensic science world. This swift development resulted in a lack of standardization and training. In his 1995 book, *"High-Technology Crime: Investigating Cases Involving Computers"*, K Rosenblatt wrote:

# Digital Forensics

Seizing, preserving, and analyzing evidence stored on a computer is the greatest forensic challenge facing law enforcement in the 1990s. Although most forensic tests, such as fingerprinting and DNA testing, are performed by specially trained experts the task of collecting and analyzing computer evidence is often assigned to patrol officers and detectives.

## *2000s: Developing standards*

Since 2000, in response to the need for standardization, various bodies and agencies have published guidelines for digital forensics. The Scientific Working Group on Digital Evidence (SWGDE) produced a 2002 paper, *"Best practices for Computer Forensics"*, this was followed, in 2005, by the publication of an ISO standard (ISO 17025, *General requirements for the competence of testing and calibration laboratories*). A European lead international treaty, the Convention on Cybercrime, came into force in 2004 with the aim of reconciling national computer crime laws, investigative techniques and international co-operation. The treaty has been signed by 43 nations (including the US, Canada, Japan, South Africa, UK and other European nations) and ratified by 16.

The issue of training also received attention. Commercial companies (often forensic software developers) began to offer certification programs and digital forensic analysis was included as a topic at the UK specialist investigator training facility, Centrex.

Since the late 1990s mobile devices have become more widely available, advancing beyond simple communication devices, and have been found to be rich forms of information, even for crime not traditionally associated with digital forensics. Despite this, digital analysis of phones has lagged behind traditional computer media, largely due to problems over the proprietary nature of devices.

Focus has also shifted onto internet crime, particularly the risk of cyber warfare and cyberterrorism. A February 2010 report by the United States Joint Forces Commandconcluded:

Through cyberspace, enemies will target industry, academia, government, as well as the military in the air, land, maritime, and space domains. In much the same way that airpower transformed the battlefield of World War II, cyberspace has fractured the physical barriers that shield a nation from attacks on its commerce and communication.

The field of digital forensics still faces unresolved issues. A 2009 paper, "Digital Forensic Research: The Good, the Bad and the Unaddressed", by Peterson and Shenoi identified a bias towards Windows operating systems in digital forensics research. In 2010 Simson Garfinkel identified issues facing digital investigations in the future, including the increasing size of digital media, the wide availability of encryption to consumers, a growing variety of operating systems and file formats, an increasing number of individuals owning multiple devices, and legal limitations on investigators. The paper also identified continued training issues, as well as the prohibitively high cost of entering the field.

## *Development of forensic tools*

During the 1980s very few specialized digital forensic tools existed, and consequently investigators often performed live analysis on media, examining computers from within the operating system using existing sysadmin tools to extract evidence. This practice carried the risk of modifying data on the disk, either inadvertently or otherwise, which led to claims of evidence tampering. A number of tools were created during the early 1990s to address the problem.

The need for such software was first recognized in 1989 at the Federal Law Enforcement Training Center, resulting in the creation of IMDUMP (by Michael White) and in 1990, SafeBack (developed by Sydex). Similar software was developed in other countries; DIBS (a hardware and software solution) was released commercially in the UK in 1991, and Rob McKemmish released *Fixed Disk Image* free to Australian law enforcement. These tools allowed examiners to create an exact copy of a piece of digital media to work on, leaving the original disk intact for verification. By the end of the 1990s, as demand for digital evidence grew more advanced commercial tools such as EnCase and FTK were developed, allowing analysts to examine copies of media without using any live forensics. More recently, a trend towards "live memory forensics" has grown resulting in the availability of tools such as WindowsSCOPE.

More recently the same progression of tool development has occurred for mobile devices; initially investigators accessed data directly on the device, but soon specialist tools such as XRY or Radio Tactics Aceso appeared.

## **COMPUTER FORENSICS**

Computer forensics (sometimes known as computer forensic science) is a branch of digital forensic science pertaining to evidence found in computers and digital storage media. The goal of computer forensics is to examine digital media in a forensically sound manner with the aim of identifying, preserving, recovering, analyzing and presenting facts and opinions about the digital information.

Although it is most often associated with the investigation of a wide variety of computer crime, computer forensics may also be used in civil proceedings. The discipline involves similar techniques and principles to data recovery, but with additional guidelines and practices designed to create a legal audit trail.

Evidence from computer forensics investigations is usually subjected to the same guidelines and practices of other digital evidence. It has been used in a number of high-profile cases and is becoming widely accepted as reliable within U.S. and European court systems.

## Overview

In the early 1980s personal computers became more accessible to consumers, leading to their increased use in criminal activity (for example, to help commit fraud). At the same time, several new "computer crimes" were recognized (such as hacking). The discipline of computer forensics emerged during this time as a method to recover and investigate digital evidence for use in court. Since then computer crime and computer related crime has grown, and has jumped 67% between 2002 and 2003. Today it is used to investigate a wide variety of crime, including child pornography, fraud, espionage, cyberstalking, murder and rape. The discipline also features in civil proceedings as a form of information gathering (for example, Electronic discovery)

Forensic techniques and expert knowledge are used to explain the current state of a *digital artifact*; such as a computer system, storage medium (e.g. hard disk or CD-ROM), an electronic document (e.g. an email message or JPEG image). The scope of a forensic analysis can vary from simple information retrieval to reconstructing a series of events.

In a 2002 book *Computer Forensics* authors Kruse and Heiser define computer forensics as involving "the preservation, identification, extraction, documentation and interpretation of computer data".

They go on to describe the discipline as "more of an art than a science", indicating that forensic methodology is backed by flexibility and extensive domain knowledge. However, while several methods can be used to extract evidence from a given computer the strategies used by law enforcement are fairly rigid and lacking the flexibility found in the civilian world.

## Use as evidence

In court, computer forensic evidence is subject to the usual requirements for digital evidence. This requires that information be authentic, reliably obtained, and admissible. Different countries have specific guidelines and practices for evidence recovery. In the United Kingdom, examiners often follow Association of Chief Police Officers guidelines that help ensure the authenticity and integrity of evidence. While voluntary, the guidelines are

widely accepted in British courts. Computer forensics has been used as evidence in criminal law since the mid-1980s, some notable examples include:
- BTK Killer: Dennis Rader was convicted of a string of serial killings that occurred over a period of sixteen years. Towards the end of this period, Rader sent letters to the police on a floppy disk. Metadata within the documents implicated an author named "Dennis" at "Christ Lutheran Church"; this evidence helped lead to Rader's arrest.
- Joseph E. Duncan III: A spreadsheet recovered from Duncan's computer contained evidence that showed him planning his crimes. Prosecutors used this to show premeditation and secure the death penalty.
- Sharon Lopatka: Hundreds of emails on Lopatka's computer lead investigators to her killer, Robert Glass.
- Corcoran Group: This case confirmed parties' duties to preserve digital evidence when litigation has commenced or is reasonably anticipated. Hard drives were analyzed by a computer forensics expert who could not find relevant emails the Defendants should have had. Though the expert found no evidence of deletion on the hard drives, evidence came out that the defendants were found to have intentionally destroyed emails, and misled and failed to disclose material facts to the plaintiffs and the court.
- Dr. Conrad Murray: Dr. Conrad Murray, the doctor of the deceased Michael Jackson, was convicted partially by digital evidence on his computer. This evidence included medical documentation showing lethal amounts of propofol.

## NETWORK FORENSICS

Network forensics is a sub-branch of digital forensics relating to the monitoring and analysis of computer network traffic for the purposes of information gathering, legal evidence, or intrusion detection. Unlike other areas of digital forensics, network investigations deal with volatile and dynamic information. Network traffic is transmitted and then lost, so network forensics is often a pro-active investigation.

Network forensics generally has two uses. The first, relating to security, involves monitoring a network for anomalous traffic and identifying intrusions. An attacker might be able to erase all log files on a compromised host; network-based evidence might therefore be the only evidence available for forensic analysis. The second form relates to law enforcement. In this case analysis of captured network traffic can include tasks such as reassembling transferred files, searching for keywords and parsing human communication such as emails or chat sessions.

Two systems are commonly used to collect network data; a brute force "catch it as you can" and a more intelligent "stop look listen" method.

## Overview

Network forensics is a comparatively new field of forensic science. The growing popularity of the Internet in homes means that computing has become network-centric and data is now available outside of disk-based digital evidence. Network forensics can be performed as a standalone investigation or alongside a computer forensics analysis (where it is often used to reveal links between digital devices or reconstruct how a crime was committed).

Marcus Ranum is credited with defining Network forensics as "the capture, recording, and analysis of network events in order to discover the source of security attacks or other problem incidents."

Compared to computer forensics, where evidence is usually preserved on disk, network data is more volatile and unpredictable. Investigators often only have material to examine if packet filters, firewalls, and intrusion detection systems were set up to anticipate breaches of security.

Systems used to collect network data for forensics use usually come in two forms:

- "Catch-it-as-you-can" - This is where all packets passing through a certain traffic point are captured and written to storage with analysis being done subsequently in batch mode. This approach requires large amounts of storage.
- "Stop, look and listen" - This is where each packet is analyzed in a rudimentary way in memory and only certain information saved for future analysis. This approach requires a faster processor to keep up with incoming traffic.

## Types

### *Ethernet*

Applying forensic methods on the Ethernet layer is done by eavesdropping bit streams with tools called monitoring tools or sniffers. The most common tool on this layer is Wireshark (formerly known as Ethereal) and tcpdump where tcpdump works mostly on unix-like operating systems. These tools collect all data on this layer and allows the user to filter for different events. With these tools, website pages, email attachments, and other network traffic can be reconstructed only if they are transmitted or received unencrypted. An advantage of collecting this data is that it is directly connected to a host. If, for example the IP address or the MAC address of a host at a certain time is known, all data sent to or from this IP or MAC address can be filtered.

To establish the connection between IP and MAC address, it is useful to take a closer look at auxiliary network protocols. The Address Resolution Protocol (ARP) tables list the MAC addresses with the corresponding IP addresses.

To collect data on this layer, the network interface card (NIC) of a host can be put into "promiscuous mode". In so doing, all traffic will be passed to the CPU, not only the traffic meant for the host.

However, if an intruder or attacker is aware that his connection might be eavesdropped, he might use encryption to secure his connection. It is almost impossible nowadays to break encryption but the fact that a suspect's connection to another host is encrypted all the time might indicate that the other host is an accomplice of the suspect.

## *TCP/IP*

On the network layer the Internet Protocol (IP) is responsible for directing the packets generated by TCP through the network (e.g., the Internet) by adding source and destination information which can be interpreted by routers all over the network. Cellular digital packet networks, like GPRS, use similar protocols like IP, so the methods described for IP work with them as well.

For the correct routing, every intermediate router must have a routing table to know where to send the packet next. These routing tables are one of the best sources of information if investigating a digital crime and trying to track down an attacker. To do this, it is necessary to follow the packets of the attacker, reverse the sending route and find the computer the packet came from (i.e., the attacker).

## *The Internet*

The internet can be a rich source of digital evidence including web browsing, email, newsgroup, synchronous chat and peer-to-peer traffic. For example, web server logs can be used to show when (or if) a suspect accessed information related to criminal activity. Email accounts can often contain useful evidence; but email headers are easily faked and, so, network forensics may be used to prove the exact origin of incriminating material. Network forensics can also be used in order to find out who is using a particular computerby extracting user account information from the network traffic.

## **DIGITAL FORENSIC PROCESS**

The digital forensic process is a recognised scientific and forensic process used in digital forensics investigations. Forensics researcher Eoghan Casey defines it as a number of steps from the original incident alert through to reporting of findings. The process is predominantly used in computer and mobile forensic investigations and consists of three steps: *acquisition, analysis* and *reporting*. Digital media seized for investigation is usually referred to as an "exhibit" in legal terminology. Investigators employ the scientific methodto recover digital evidence to support or disprove a hypothesis, either for a court of law or in civil proceedings.

## Personnel

The stages of the digital forensics process require differing specialist training and knowledge, there are two rough levels of personnel:

### *Digital forensic technician*

Technicians may gather or process evidence at crime scenes, in the field of digital forensics training is needed on the correct handling of technology (for example to preserve the evidence). Technicians may be required to carry out "Live analysis" of evidence - various tools to simplify this procedure have been produced, most notably Microsoft'sCOFEE.

### *Digital Evidence Examiners*

Examiners specialize in one area of digital evidence; either at a broad level (i.e. computer or network forensics etc.) or as a sub-specialist (i.e. image analysis)

## Seizure

Prior to the actual examination digital media will be seized. In criminal cases this will often be performed by law enforcement personnel trained as technicians to ensure the preservation of evidence. In civil matters it will usually be a company officer, often untrained. Various laws cover the seizure of material. In criminal matters law related tosearch warrants is applicable. In civil proceedings the assumption is that a company is able to investigate their own equipment without a warrant, so long as the privacy and human rights of employees are observed.

## Acquisition

Once exhibits have been seized an exact sector level duplicate (or "forensic duplicate") of the media is created, usually via a write blocking device, a process referred to as *Imaging* or *Acquisition*. The duplicate is created using a hard-drive duplicator or software imaging tools such as DCFLdd, IXimager, Guymager, TrueBack, EnCase, FTK Imager or FDAS. The original drive is then returned to secure storage to prevent tampering. The acquired image is verified by using the SHA-1 or MD5 hash functions. At critical points throughout the analysis, the media is verified again, known as "hashing", to ensure that the evidence is still in its original state.

## Analysis

After acquisition the contents of (the HDD) image files are analysed to identify evidence that either supports or contradicts a hypothesis or for signs of tampering (to hide data). In 2002 the *International Journal of Digital Evidence* referred to this stage as "an in-depth systematic search of evidence related to the suspected crime".

By contrast Brian Carrier, in 2006, describes a more "intuitive procedure" in which obvious evidence is first identified after which "exhaustive searches are conducted to start filling in the holes"

During the analysis an investigator usually recovers evidence material using a number of different methodologies (and tools), often beginning with recovery of deleted material. Examiners use specialist tools (EnCase, ILOOKIX, FTK, etc.) to aid with viewing and recovering data. The type of data recovered varies depending on the investigation; but examples include email, chat logs, images, internet history or documents. The data can be recovered from accessible disk space, deleted (unallocated) space or from within operating system cache files.

Various types of techniques are used to recover evidence, usually involving some form of keyword searching within the acquired image file; either to identify matches to relevant phrases or to parse out known file types. Certain files (such as graphic images) have a specific set of bytes which identify the start and end of a file, if identified a deleted file can be reconstructed. Many forensic tools use hash signatures to identify notable files or to exclude known (benign) ones; acquired data is hashed and compared to pre-compiled lists such as the *Reference Data Set* (RDS) from the National Software Reference Library

On most media types including standard magnetic hard disks, once data has been securely deleted it can never be recovered. SSD Drives are specifically of interest from a forensics viewpoint, because even after a secure-erase operation some of the data that was intended to be secure-erased persists on the drive.

Once evidence is recovered the information is analysed to reconstruct events or actions and to reach conclusions, work that can often be performed by less specialist staff.Digital investigators, particularly in criminal investigations, have to ensure that conclusions are based upon data and their own expert knowledge. In the US, for example, Federal Rules of Evidence state that a qualified expert may testify "in the form of an opinion or otherwise" so long as: (1) the testimony is based upon sufficient facts or data, (2) the testimony is the product of reliable principles and methods, and (3) the witness has applied the principles and methods reliably to the facts of the case.

# Digital Forensics

## Reporting

When an investigation is completed the information is often reported in a form suitable for non-technical individuals. Reports may also include audit information and other meta-documentation.

When completed reports are usually passed to those commissioning the investigation, such as law enforcement (for criminal cases) or the employing company (in civil cases), who will then decide whether to use the evidence in court. Generally, for a criminal court, the report package will consist of a written expert conclusion of the evidence as well as the evidence itself (often presented on digital media).

## FORENSIC PROCESS

A digital forensic investigation commonly consists of 3 stages: acquisition or imaging of exhibits, analysis, and reporting. Ideally acquisition involves capturing an image of the computer's volatile memory (RAM) and creating an exact sector level duplicate (or "forensic duplicate") of the media, often using a write blocking device to prevent modification of the original.

However, the growth in size of storage media and developments such as cloud computing have led to more use of 'live' acquisitions whereby a 'logical' copy of the data is acquired rather than a complete image of the physical storage device. Both acquired image (or logical copy) and original media/data are hashed (using an algorithm such as SHA-1 or MD5) and the values compared to verify the copy is accurate.

During the analysis phase an investigator recovers evidence material using a number of different methodologies and tools. In 2002, an article in the *International Journal of Digital Evidence* referred to this step as "an in-depth systematic search of evidence related to the suspected crime." In 2006, forensics researcher Brian Carrier described an "intuitive procedure" in which obvious evidence is first identified and then "exhaustive searches are conducted to start filling in the holes."

The actual process of analysis can vary between investigations, but common methodologies include conducting keyword searches across the digital media (within files as well as unallocated and slack space), recovering deleted files and extraction of registry information (for example to list user accounts, or attached USB devices).

The evidence recovered is analysed to reconstruct events or actions and to reach conclusions, work that can often be performed by less specialised staff. When an investigation is complete the data is presented, usually in the form of a written report, in lay persons' terms.

## Application

Digital forensics is commonly used in both criminal law and private investigation. Traditionally it has been associated with criminal law, where evidence is collected to support or oppose a hypothesis before the courts. As with other areas of forensics this is often as part of a wider investigation spanning a number of disciplines. In some cases the collected evidence is used as a form of intelligence gathering, used for other purposes than court proceedings (for example to locate, identify or halt other crimes). As a result, intelligence gathering is sometimes held to a less strict forensic standard.

In civil litigation or corporate matters digital forensics forms part of the electronic discovery (or eDiscovery) process. Forensic procedures are similar to those used in criminal investigations, often with different legal requirements and limitations. Outside of the courts digital forensics can form a part of internal corporate investigations. A common example might be following unauthorized network intrusion. A specialist forensic examination into the nature and extent of the attack is performed as a damage limitation exercise. Both to establish the extent of any intrusion and in an attempt to identify the attacker. Such attacks were commonly conducted over phone lines during the 1980s, but in the modern era are usually propagated over the Internet.

The main focus of digital forensics investigations is to recover objective evidence of a criminal activity (termed actus reus in legal parlance). However, the diverse range of data held in digital devices can help with other areas of inquiry.

### *Attribution*

Meta data and other logs can be used to attribute actions to an individual. For example, personal documents on a computer drive might identify its owner.

### *Alibis and statements*

Information provided by those involved can be cross checked with digital evidence. For example, during the investigation into the Soham murders the offender's alibi was disproved when mobile phone records of the person he claimed to be with showed she was out of town at the time.

### *Intent*

As well as finding objective evidence of a crime being committed, investigations can also be used to prove the intent (known by the legal term mens rea). For example, the Internet history of convicted killer Neil Entwistle included references to a site discussing *How to kill people*.

## Evaluation of source

File artifacts and meta-data can be used to identify the origin of a particular piece of data; for example, older versions of Microsoft Word embedded a Global Unique Identifer into files which identified the computer it had been created on. Proving whether a file was produced on the digital device being examined or obtained from elsewhere (e.g., the Internet) can be very important.

## Document authentication

Related to "Evaluation of source," meta data associated with digital documents can be easily modified (for example, by changing the computer clock you can affect the creation date of a file). Document authentication relates to detecting and identifying falsification of such details.

## Limitations

One major limitation to a forensic investigation is the use of encryption; this disrupts initial examination where pertinent evidence might be located using keywords. Laws to compel individuals to disclose encryption keys are still relatively new and controversial.

## Legal considerations

The examination of digital media is covered by national and international legislation. For civil investigations, in particular, laws may restrict the abilities of analysts to undertake examinations. Restrictions against network monitoring, or reading of personal communications often exist. During criminal investigation, national laws restrict how much information can be seized.

For example, in the United Kingdom seizure of evidence by law enforcement is governed by the PACE act. During its existence early in the field, the "International Organization on Computer Evidence" (IOCE) was one agency that worked to establish compatible international standards for the seizure of evidence.

In the UK the same laws covering computer crime can also affect forensic investigators. The 1990 computer misuse act legislates against unauthorised access to computer material; this is a particular concern for civil investigators who have more limitations than law enforcement.

An individuals right to privacy is one area of digital forensics which is still largely undecided by courts. The US Electronic Communications Privacy Act places limitations on the ability of law enforcement or civil investigators to intercept and access evidence. The act makes a distinction between stored communication (e.g. email archives) and transmitted communication (such as VOIP). The latter, being considered more of a privacy invasion, is harder to obtain a warrant for. The ECPA also affects the ability of companies to investigate the computers and communications of their employees, an aspect that is still under debate as to the extent to which a company can perform such monitoring.

Article 5 of the European Convention on Human Rights asserts similar privacy limitations to the ECPA and limits the processing and sharing of personal data both within the EU and with external countries. The ability of UK law enforcement to conduct digital forensics investigations is legislated by the Regulation of Investigatory Powers Act.

## Digital evidence

When used in a court of law digital evidence falls under the same legal guidelines as other forms of evidence; courts do not usually require more stringent guidelines. In the United States the Federal Rules of Evidence are used to evaluate the admissibility of digital evidence, the United Kingdom PACE and Civil Evidence acts have similar guidelines and many other countries have their own laws. US federal laws restrict seizures to items with only obvious evidential value. This is acknowledged as not always being possible to establish with digital media prior to an examination.

Laws dealing with digital evidence are concerned with two issues: integrity and authenticity. Integrity is ensuring that the act of seizing and acquiring digital media does not modify the evidence (either the original or the copy). Authenticity refers to the ability to confirm the integrity of information; for example that the imaged media matches the original evidence. The ease with which digital media can be modified means that documenting the chain of custody from the crime scene, through analysis and, ultimately, to the court, (a form of audit trail) is important to establish the authenticity of evidence.

Attorneys have argued that because digital evidence can theoretically be altered it undermines the reliability of the evidence. US judges are beginning to reject this theory, in the case *US v. Bonallo* the court ruled that "the fact that it is possible to alter data contained in a computer is plainly insufficient to establish untrustworthiness." In the United Kingdom guidelines such as those issued by ACPO are followed to help document the authenticity and integrity of evidence.

Digital investigators, particularly in criminal investigations, have to ensure that conclusions are based upon factual evidence and their own expert knowledge. In the US, for example, Federal Rules of Evidence state that a qualified expert may testify "in the form of an opinion or otherwise" so long as: (1) the testimony is based upon sufficient facts or data, (2) the testimony is the product of reliable principles and methods, and (3) the witness has applied the principles and methods reliably to the facts of the case.

The sub-branches of digital forensics may each have their own specific guidelines for the conduct of investigations and the handling of evidence. For example, mobile phones may be required to be placed in a Faraday shield during seizure or acquisition to prevent further radio traffic to the device. In the UK forensic examination of computers in criminal matters is subject to ACPO guidelines. There are also international approaches to providing guidance on how to handle electronic evidence. The "Electronic Evidence Guide" by the Council of Europe offers a framework for law enforcement and judicial authorities in countries who seek to set up or enhance their own guidelines for the identification and handling of electronic evidence.

## *Investigative tools*

The admissibility of digital evidence relies on the tools used to extract it. In the US, forensic tools are subjected to the Daubert standard, where the judge is responsible for ensuring that the processes and software used were acceptable. In a 2003 paper Brian Carrier argued that the Daubert guidelines required the code of forensic tools to be published and peer reviewed. He concluded that "open source tools may more clearly and comprehensively meet the guideline requirements than would closed source tools."

## Branches

Digital forensics includes several sub-branches relating to the investigation of various types of devices, media or artifacts.

## *Computer forensics*

The goal of computer forensics is to explain the current state of a digital artifact; such as a computer system, storage medium or electronic document. The discipline usually covers computers, embedded systems (digital devices with rudimentary computing power and onboard memory) and static memory (such as USB pen drives).

Computer forensics can deal with a broad range of information; from logs (such as internet history) through to the actual files on the drive. In 2007 prosecutors used aspreadsheet recovered from the computer of Joseph E. Duncan III to show premeditation and secure the death penalty. Sharon Lopatka's killer was identified in 2006 after email messages from him detailing torture and death fantasies were found on her computer.

## *Mobile device forensics*

Mobile device forensics is a sub-branch of digital forensics relating to recovery of digital evidence or data from a mobile device. It differs from Computer forensics in that a mobile device will have an inbuilt communication system (e.g. GSM) and, usually, proprietary storage mechanisms. Investigations usually focus on simple data such as call data and communications (SMS/

Email) rather than in-depth recovery of deleted data. SMS data from a mobile device investigation helped to exonerate Patrick Lumumba in the murder of Meredith Kercher. Mobile devices are also useful for providing location information; either from inbuilt gps/location tracking or via cell site logs, which track the devices within their range. Such information was used to track down the kidnappers of Thomas Onofri in 2006.

## Network forensics

Network forensics is concerned with the monitoring and analysis of computer network traffic, both local and WAN/internet, for the purposes of information gathering, evidence collection, or intrusion detection.

Traffic is usually intercepted at the packet level, and either stored for later analysis or filtered in real-time. Unlike other areas of digital forensics network data is often volatile and rarely logged, making the discipline often reactionary.

In 2000 the FBI lured computer hackers Aleksey Ivanov and Gorshkov to the United States for a fake job interview. By monitoring network traffic from the pair's computers, the FBI identified passwords allowing them to collect evidence directly from Russian-based computers.

## Forensic data analysis

Forensic Data Analysis is a branch of digital forensics. It examines structured data with the aim to discover and analyse patterns of fraudulent activities resulting from financial crime.

## Database forensics

Database forensics is a branch of digital forensics relating to the forensic study of databases and their metadata. Investigations use database contents, log files and in-RAMdata to build a timeline or recover relevant information.

## **Education and Research**

Academic centre of education and research in forensic sciences:

North America: Penn State University offers Security and Risk Analysis Major, Master of Professional Studies in Information Sciences, Master of Professional Studies in Homeland Security, and Ph.D. in Information Sciences and Technology in the digital forensics area.

## **MOBILE DEVICE FORENSICS**

Mobile device forensics is a branch of digital forensics relating to recovery of digital evidence or data from a mobile device underforensically sound conditions. The phrase *mobile device* usually refers to mobile phones; however, it can also relate to any digital device that has both internal memory and communication ability, including PDA devices, GPS devices and tablet computers.

The use of phones in crime was widely recognised for some years, but the forensic study of mobile devices is a relatively new field, dating from the early 2000s.

A proliferation of phones (particularly smartphones) on the consumer market caused a demand for forensic examination of the devices, which could not be met by existing computer forensics techniques.

Mobile devices can be used to save several types of personal information such as contacts, photos, calendars and notes, SMS and MMSmessages. Smartphones may additionally contain video, email, web browsing information, location information, and social networking messages and contacts.

There is growing need for mobile forensics due to several reasons and some of the prominent reasons are:
- Use of mobile phones to store and transmit personal and corporate information
- Use of mobile phones in online transactions
- Law enforcement, criminals and mobile phone devices

Mobile device forensics can be particularly challenging on a number of levels:

Evidential and technical challenges exist. for example, cell site analysis following from the use of a mobile phone usage coverage, is not an exact science. Consequently, whilst it is possible to determine roughly the cell site zone from which a call was made or received, it is not yet possible to say with any degree of certainty, that a mobile phone call emanated from a specific location e.g. a residential address.
- To remain competitive, original equipment manufacturers frequently change mobile phone form factors, operating system file structures, data storage, services, peripherals, and even pin connectors and cables. As a result, forensic examiners must use a different forensic process compared to computer forensics.
- Storage capacity continues to grow thanks to demand for more powerful "mini computer" type devices.
- Not only the types of data but also the way mobile devices are used constantly evolve.
- Hibernation behaviour in which processes are suspended when the device is powered off or idle but at the same time, remaining active.

As a result of these challenges, a wide variety of tools exist to extract evidence from mobile devices; no one tool or method can acquire all the evidence from all devices. It is therefore recommended that forensic examiners, especially those wishing to qualify as expert witnesses in court, undergo extensive training in order to understand how each tool and method acquires evidence; how it maintains standards for forensic soundness; and how it meets legal requirements such as the Daubert standard or Frye standard.

## History

As a field of study forensic examination of mobile devices dates from the late 1990s and early 2000s. The role of mobile phones in crime had long been recognized by law enforcement. With the increased availability of such devices on the consumer market and the wider array of communication platforms they support (e.g. email, web browsing) demand for forensic examination grew.

Early efforts to examine mobile devices used similar techniques to the first computer forensics investigations: analysing phone contents directly via the screen and photographing important content. However, this proved to be a time-consuming process, and as the number of mobile devices began to increase, investigators called for more efficient means of extracting data. Enterprising mobile forensic examiners sometimes used cell phone or PDA synchronization software to "back up" device data to a forensic computer for imaging, or sometimes, simply performed computer forensics on the hard drive of a suspect computer where data had been synchronized. However, this type of software could write to the phone as well as reading it, and could not retrieve deleted data.

Some forensic examiners found that they could retrieve even deleted data using "flasher" or "twister" boxes, tools developed by OEMs to "flash" a phone's memory for debugging or updating. However, flasher boxes are invasive and can change data; can be complicated to use; and, because they are not developed as forensic tools, perform neither hash verifications nor (in most cases) audit trails. For physical forensic examinations, therefore, better alternatives remained necessary.

To meet these demands, commercial tools appeared which allowed examiners to recover phone memory with minimal disruption and analyse it separately. Over time these commercial techniques have developed further and the recovery of deleted data from proprietary mobile devices has become possible with some specialist tools.

Moreover, commercial tools have even automated much of the extraction process, rendering it possible even for minimally trained first responders—who currently are much more likely to encounter suspects with mobile devices in their possession, compared to computers—to perform basic extractions for triage and data preview purposes.

# Digital Forensics

## Professional Applications

Mobile device forensics is best known for its application to law enforcement investigations, but it is also useful for military intelligence, corporate investigations, private investigations, criminal and civil defense, and electronic discovery.

## Types of evidence

As mobile device technology advances, the amount and types of data that can be found on a mobile device is constantly increasing. Evidence that can be potentially recovered from a mobile phone may come from several different sources, including handset memory, SIM card, and attached memory cards such as SD cards.

Traditionally mobile phone forensics has been associated with recovering SMS and MMS messaging, as well as call logs, contact lists and phone IMEI/ESN information. However, newer generations of smartphones also include wider varieties of information; from web browsing, Wireless network settings, geolocation information (includinggeotags contained within image metadata), e-mail and other forms of rich internet media, including important data—such as social networking service posts and contacts—now retained on smartphone 'apps'.

### *Internal memory*

Nowadays mostly flash memory consisting of NAND or NOR types are used for mobile devices.

### *External memory*

External memory devices are SIM cards, SD cards (commonly found within GPS devices as well as mobile phones), MMC cards, CF cards, and the Memory Stick.

### *Service provider logs*

Although not technically part of mobile device forensics, the call detail records (and occasionally, text messages) from wireless carriers often serve as "back up" evidence obtained after the mobile phone has been seized. These are useful when the call history and/or text messages have been deleted from the phone, or when location-based services are not turned on. Call detail records and cell site (tower) dumps can show the phone owner's location, and whether they were stationary or moving (i.e., whether the phone's signal bounced off the same side of a single tower, or different sides of multiple towers along a particular path of travel). Carrier data and device data together can be used to corroborate information from other sources, for instance, video surveillance footage or eyewitness accounts; or to determine the general location where a non-geotagged image or video was taken.

The European Union requires its member countries to retain certain telecommunications data for use in investigations. This includes data on calls made and retrieved. The location of a mobile phone can be determined and this geographical data must also be retained. In the United States, however, no such requirement exists, and no standards govern how long carriers should retain data or even what they must retain. For example, text messages may be retained only for a week or two, while call logs may be retained anywhere from a few weeks to several months. To reduce the risk of evidence being lost, law enforcement agents must submit a preservation letter to the carrier, which they then must back up with a search warrant.

# 3

# Forensic Sciences

The forensic sciences have played a key role in criminal investigations for many years. Recently, there has been increased attention on the forensic sciences by law enforcement, prosecutors, and the general public. Particularly in high profile cases, intense media coverage concerning evidence issues and the work of crime laboratories has served to heighten this interest.

In the past two decades, there have been tremendous technological advances in the laboratory testing of forensic samples. There have also been a number of improvements in the identification and collection of evidence at the crime scene, through innovative processing and evidence collection methods. Together, these advances allow for a greater probability of successful recovery and analysis of evidence than was previously possible. There is also growing recognition by criminal justice professionals of the wider scope of forensic techniques and available tests.

The field of forensic deoxyribonucleic acid (DNA) analysis and the legislation that allows DNA testing on a broader number of offenders has made some of the more remarkable advances. DNA testing now allows much smaller samples of biological material to be analysed and the results to be more discriminating. DNA testing of forensic crime scene samples can now be compared against a database of known offenders and other unsolved crimes.

Forensic laboratories have developed advanced analytical techniques through the use of computer technology. Systems such as the Combined DNA Index System (CODIS), various Automated Fingerprint Identification Systems (AFIS), and the National Integrated Ballistics Identification Network (NIBIN), were identified by the symposium as beneficial to serial murder investigations, by providing links between previously unrelated cases.

CODIS is a national automated DNA information processing and telecommunications system that was developed to link biological evidence (DNA) in criminal cases, between various jurisdictions around the United

States. Samples in CODIS include DNA profiles obtained from persons convicted of designated crimes, DNA profiles obtained from crime scenes, DNA profiles from unidentified human remains, and DNA from voluntary samples taken from families of missing persons.

The CODIS data bank of these samples is comprised of three different indices or levels: the National DNA Index System (NDIS), the State DNA Index System (SDIS), and the Local DNA Index System (LDIS).

What is important for law enforcement to understand is that the information contained at the LDIS and SDIS levels may not automatically be sent to, or searched against, the NDIS level. There are different legislation requirements for inclusion into NDIS, than to LDIS or SDIS, and not all LDIS and SDIS profiles are sent to NDIS. Even when NDIS is queried, individual SDIS data banks may not be queried. Therefore, when dealing with a serial murder case, investigators need to contact their LDIS or SDIS level representatives to ensure that in addition to the NDIS databank, samples are compared in the individual SDIS data banks of each state that is of investigative interest. In cases where there is only a partial DNA profile, a national "keyboard" search can be requested through the NDIS custodian, CODIS Unit, FBI Laboratory.

AFIS is an electronic databank that compares unidentified latent and patent fingerprints to the known fingerprint file. There have been a variety of local AFIS systems in use since the 1980s. In 1999, the FBI's Integrated Automated Fingerprint Identification System, or IAFIS, became operational. IAFIS is designated as the national repository of criminal histories, fingerprints, and photographs of criminal subjects in the United States. It also contains fingerprints and information on military and civilian federal employees. IAFIS provides positive identification through comparisons of individuals based on the submission of fingerprint data, through both ten-print fingerprint cards and latent fingerprints.

Some of the earlier AFIS systems were not compatible with the IAFIS system, and as a result, those earlier latent fingerprints may not be included in IAFIS. This becomes an issue in serial murder cases, when the offender committed offences prior to the inception of IAFIS, as latent fingerprints from those earlier crimes will not be searchable. If there is a possibility the offender committed early crimes, the early AFIS systems need to be queried independently. Consultation with laboratory fingerprint experts may be necessary in order to establish what AFIS systems exist, which are interoperable, and the protocols required to query each system.

NIBIN is a national databank of both projectile and cartridge information. NIBIN is the integration of two previous systems: the FBI's Drugfire cartridge case imaging system and the Bureau of Alcohol, Tobacco, Firearms and

Explosives' (ATF) Integrated Ballistic Identification System (IBIS). NIBIN is an imaging system that allows both bullets and cartridges recovered from a crime scene to be compared electronically against other bullets and cartridges recovered from previous crime scenes, in an effort to link previously unrelated cases. The system can search by geographic area or nationwide, depending upon the course of the investigation. ATF is maintaining the new system in over 75 locations, across the United States.

When conducting serial murder investigations, it is important for investigators to promptly seek guidance from appropriate forensic database experts. Such experts can provide information regarding what limitations exist and what additional queries can be made of the systems, to obtain additional investigative information.

Another area in which forensic science can play an important role is in the recovery and examination of trace evidence. Trace evidence is described as small, often microscopic material. It commonly includes hair and fiber evidence but may encompass almost any substance or material. Trace evidence may provide important lead information pertaining to offender characteristics, vehicle and tire descriptors, and environmental clues that relate to killing scenes and modes of transportation used to move bodies.

A skilled trace evidence examiner can compare the trace evidence from all of the victims in a serial murder case, in an effort to identify evidence common to all of the victims. This trace evidence will reflect a "common environment" with which all of the victims were in contact.

This common environment will repeat in objects in the serial offender's world, such as his vehicles and/or residence. This can demonstrate that all of the victims had contact with the offender at the same location(s).

Attends at the Serial Murder Symposium universally acknowledged that serial murder cases present unique circumstances and concerns, particularly when multiple investigative jurisdictions are involved. In serial murder cases, crime scenes may occur in different law enforcement jurisdictions, each of whom may possess varying resources and abilities to process crime scenes. In some cases, agencies submit evidence to different laboratories, even though those agencies are located adjacent to one another. These issues degrade the ability of law enforcement to consistently collect evidence from a murder series. This may prevent identifying a serial killer or forensically linking previously unrelated cases to a common offender.

Attends identified a number of forensic issues facing the law enforcement community in serial murder investigations and made the following suggestions:
- Once a series is identified, the same crime scene personnel should be utilized at related scenes to promote consistency in evidence identification and collection. Search personnel should follow established

sterilization procedures to ensure there is no cross-contamination between the various crime scenes.
- Cross-contamination should be proactively prevented by using different personnel to process crime scenes than those used to collect known sample evidence from potential suspects.
- Documentation among the law enforcement agencies should be standardized to ensure continuity between separate cases.
- Aerial photographs of every murder crime scene, as well as the accompanying ancillary scenes, should be taken. Aerial photographs clearly depict the geography of the area and demonstrate the physical relationships and the distances between the crime scenes. They also identify potential routes of ingress and egress to the area.
- The number of laboratories and experts involved in serial murder investigations should be limited to properly certified facilities and personnel. Ideally, all evidence should be examined by a single crime laboratory, and that lab should utilize only one expert per discipline. If this is not possible, establish lines of communication between laboratories to ensure the sharing of pertinent information related to the investigation.
- Priority status for laboratory examinations should be obtained to ensure a quick turn around on test results.
- When consulting with forensic scientists, investigators should prioritize forensic examinations based upon their potential investigative value. In addition, forensic scientists should be consulted frequently to identify alternative sampling and/or testing that may lead to successful case resolution.
- Forensic testimony should be limited to what is needed for successful prosecution. Utilization of charts, graphs, or other appropriate audiovisual aides showing forensic linkages will clearly and succinctly convey the facts of the cases.
- When necessary, investigators should seek independent, secondary reviews of laboratory results. This may be somewhat problematic, since there are crime laboratories that will not duplicate forensic examinations. However, exceptions are sometimes made to this policy on a case-by-case basis.

## HISTORY

### Early methods

The ancient world lacked standardized forensic practices, which aided criminals in escaping punishment. Criminal investigations and trials heavily relied on forced confessions and witness testimony. However, ancient sources do contain several accounts of techniques that foreshadow concepts in forensic science that were developed centuries later. For instance, Archimedes (287–212 BC) invented a method for determining the volume of an object with an irregular shape. According to Vitruvius, a votive crown for a temple had been made for King Hiero II, who had supplied the pure gold to be used, and Archimedes was asked to determine whether some silver had been substituted by the dishonest goldsmith. Archimedes had to solve the problem without damaging the crown, so he could not melt it down into a regularly shaped body in order to calculate its density. Instead he used the law of displacement to prove that the goldsmith had taken some of the gold and substituted silver instead.

The first written account of using medicine and entomology to solve criminal cases is attributed to the book of *Xi Yuan Lu* (translated as *Washing Away of Wrongs*), written in China by Song Ci (1186–1249) in 1248, during the Song Dynasty. In one of the accounts, the case of a person murdered with a sickle was solved by an investigator who instructed everyone to bring his sickle to one location. (He realized it was a sickle by testing various blades on an animal carcass and comparing the wound.) Flies, attracted by the smell of blood, eventually gathered on a single sickle. In light of this, the murderer confessed. The book also offered advice on how to distinguish between a drowning(water in the lungs) and strangulation (broken neck cartilage), along with other evidence from examining corpses on determining if a death was caused by murder, suicide or an accident.

Methods from around the world involved saliva and examination of the mouth and tongue to determine innocence or guilt, as a precursor to thePolygraph test. In ancient India, some suspects were made to fill their mouths with dried rice and spit it back out. Similarly, in Ancient China, those accused of a crime would have rice powder placed in their mouths. In ancient middle-eastern cultures, the accused were made to lick hot metal rods briefly. It is thought that these tests had some validity since a guilty person would produce less saliva and thus have a drier mouth; the accused would be considered guilty if rice was sticking to their mouths in abundance or if their tongues were severely burned due to lack of shielding from saliva.

## Origins of forensic science

In 16th-century Europe, medical practitioners in army and university settings began to gather information on the cause and manner of death. Ambroise Paré, a French army surgeon, systematically studied the effects of violent death on internal organs. Two Italiansurgeons, Fortunato Fidelis and Paolo Zacchia, laid the foundation of modern pathology by studying changes that occurred in the structure of the body as the result of disease. In the late 18th century, writings on these topics began to appear. These included *A Treatise on Forensic Medicine and Public Health* by the French physician Francois Immanuele Fodéré and *The Complete System of Police Medicine* by the German medical expert Johann Peter Frank.

As the rational values of the Enlightenment era increasingly permeated society in the 18th century, criminal investigation became a more evidence-based, rational procedure " the use of torture to force confessions was curtailed, and belief in witchcraft and other powers of the occult largely ceased to influence the court's decisions. Two examples of English forensic science in individual legal proceedings demonstrate the increasing use of logic and procedure in criminal investigations at the time. In 1784, in Lancaster, John Toms was tried and convicted for murdering Edward Culshaw with a pistol. When the dead body of Culshaw was examined, a pistol wad (crushed paper used to secure powder and balls in the muzzle) found in his head wound matched perfectly with a torn newspaper found in Toms's pocket, leading to the conviction.

In Warwick in 1816, a farm labourer was tried and convicted of the murder of a young maidservant. She had been drowned in a shallow pool and bore the marks of violent assault. The police found footprints and an impression from corduroy cloth with a sewn patch in the damp earth near the pool. There were also scattered grains of wheat and chaff. The breeches of a farm labourer who had been threshing wheat nearby were examined and corresponded exactly to the impression in the earth near the pool.

## Toxicology and ballistics

A method for detecting arsenious oxide, simple arsenic, in corpses was devised in 1773 by the Swedish chemist Carl Wilhelm Scheele. His work was expanded, in 1806, by German chemist Valentin Ross, who learned to detect the poison in the walls of a victim's stomach.

James Marsh was the first to apply this new science to the art of forensics. He was called by the prosecution in a murder trial to give evidence as a chemist in 1832. The defendant, John Bodle, was accused of poisoning his grandfather with arsenic-laced coffee. Marsh performed the standard test by mixing a suspected sample with hydrogen sulfide and hydrochloric acid. While he was able to detect arsenic as yellow arsenic trisulfide, when it was shown to the jury it had deteriorated, allowing the suspect to be acquitted due to reasonable doubt.

Annoyed by this, Marsh developed a much better test. He combined a sample containing arsenic with sulfuric acid and arsenic-free zinc, resulting in arsine gas. The gas was ignited, and it decomposed to pure metallic arsenic, which, when passed to a cold surface, would appear as a silvery-black deposit. So sensitive was the test, known formally as the Marsh test, that it could detect as little as one-fiftieth of a milligram of arsenic. He first described this test in *The Edinburgh Philosophical Journal* in 1836.

Henry Goddard at Scotland Yard pioneered the use of bullet comparison in 1835. He noticed a flaw in the bullet that killed the victim and was able to trace this back to the mold that was used in the manufacturing process.

## Anthropometry

The French police officer Alphonse Bertillon was the first to apply the anthropological technique of anthropometry to law enforcement, thereby creating an identification system based on physical measurements. Before that time, criminals could only be identified by name or photograph.Dissatisfied with the *ad hoc* methods used to identify captured criminals in France in the 1870s, he began his work on developing a reliable system of anthropometrics for human classification.

Bertillon created many other forensics techniques, including forensic document examination, the use of galvanoplastic compounds to preservefootprints, ballistics, and the dynamometer, used to determine the degree of force used in breaking and entering. Although his central methods were soon to be supplanted by fingerprinting, "his other contributions like the mug shot and the systematization of crime-scene photography remain in place to this day."

## Fingerprints

Sir William Herschel was one of the first to advocate the use of fingerprinting in the identification of criminal suspects. While working for the Indian Civil Service, he began to use thumbprints on documents as a security measure to prevent the then-rampant repudiation of signatures in 1858.

In 1877 at Hooghly (near Calcutta), Herschel instituted the use of fingerprints on contracts and deeds, and he registered government pensioners' fingerprints to prevent the collection of money by relatives after a pensioner's death.

In 1880, Dr. Henry Faulds, a Scottish surgeon in a Tokyo hospital, published his first paper on the subject in the scientific journal *Nature*, discussing the usefulness of fingerprints for identification and proposing a method to record them with printing ink. He established their first classification and was also the first to identify fingerprints left on a vial. Returning to the UK in 1886, he offered the concept to the Metropolitan Police in London, but it was dismissed at that time.

Faulds wrote to Charles Darwin with a description of his method, but, too old and ill to work on it, Darwin gave the information to his cousin, Francis Galton, who was interested in anthropology. Having been thus inspired to study fingerprints for ten years, Galton published a detailed statistical model of fingerprint analysis and identification and encouraged its use in forensic science in his book *Finger Prints*. He had calculated that the chance of a "false positive" (two different individuals having the same fingerprints) was about 1 in 64 billion.

Juan Vucetich, an Argentine chief police officer, created the first method of recording the fingerprints of individuals on file. In 1892, after studying Galton's pattern types, Vucetich set up the world's first fingerprint bureau. In that same year, Francisca Rojas of Necochea was found in a house with neck injuries whilst her two sons were found dead with their throats cut. Rojas accused a neighbour, but despite brutal interrogation, this neighbour would not confess to the crimes.

Inspector Alvarez, a colleague of Vucetich, went to the scene and found a bloody thumb mark on a door. When it was compared with Rojas' prints, it was found to be identical with her right thumb. She then confessed to the murder of her sons.

A Fingerprint Bureau was established in Calcutta (Kolkata), India, in 1897, after the Council of the Governor General approved a committee report that fingerprints should be used for the classification of criminal records. Working in the Calcutta Anthropometric Bureau, before it became the Fingerprint Bureau, were Azizul Haque and Hem Chandra Bose. Haque and Bose were Indian fingerprint experts who have been credited with the primary development of a fingerprint classification system eventually named after their supervisor, Sir Edward Richard Henry. The Henry Classification System, co-devised by Haque and Bose, was accepted in England and Wales when the first United Kingdom Fingerprint Bureau was founded in Scotland Yard, the Metropolitan Police headquarters, London, in 1901. Sir Edward Richard Henry subsequently achieved improvements in dactyloscopy.

In the United States, Dr. Henry P. DeForrest used fingerprinting in the New York Civil Service in 1902, and by December 1905, New York City Police Department Deputy Commissioner Joseph A. Faurot, an expert in the Bertillon system and a fingerprint advocate at Police Headquarters, introduced the fingerprinting of criminals to the United States.

## DNA

DNA fingerprinting was first used in 1984. It was discovered by Sir Alec Jefferys who realized that variation in the genetic code could be used to identify individuals and to tell individuals apart from one another. The first application of DNA profiles was used by Jefferys in a double murder mystery in a small England town called Narborough, Leicestershire in 1983. A 15-year-old school girl by the name of Lynda Mann was raped and murdered in Carlton Hayes psychiatric hospital. The police did not find a suspect but were able to obtain a semen sample.

In 1986, Dawn Ashworth, 15 years old, was also raped and strangled in a nearby village of Enderby. Forensic evidence showed that both killers had the same blood type. Richard Buckland became the suspect because he worked at Carlton Hayes psychiatric hospital, had been spotted near Dawn Ashworth's murder scene and knew unreleased details about the body. He later confessed to Dawn's murder but not Lynda's. Sir Alec Jefferys was brought into case to analyze the semen samples. He concluded that there was no match between the samples and Buckland, who became the first person to be exonerated using DNA. Jefferys confirmed that the DNA profiles were identical for the two murder semen samples. To find the perpetrator, DNA from entire male population, more than 4,000 aged from 17 to 34, in town was collected. They all were compared to semen samples from the crime.

A friend of Colin Pitchfork was heard saying that he had given his sample to the police claiming to be Colin. Colin Pitchfork was arrested in 1987 and it was found that his DNA profile matched the semen samples from the murder.

Because of this case, DNA databases came into being. There is the national (FBI) and international databases as well as the European countries (ENFSI). These searchable databases are used to match crime scene DNA profiles to those already in database.

## *Maturation*

By the turn of the 20th century, the science of forensics had become largely established in the sphere of criminal investigation. Scientific and surgical investigation was widely employed by the Metropolitan Police during their pursuit of the mysterious Jack the Ripper, who had killed a series of prostitutes in the 1880s. This case is a watershed in the application of forensic science. Large teams of policemen conducted house-to-house inquiries throughout Whitechapel. Forensic material was collected and examined. Suspects were identified, traced and either examined more closely or eliminated from the inquiry. Police work follows the same pattern today. Over 2000 people were interviewed, "upwards of 300" people were investigated, and 80 people were detained.

The investigation was initially conducted by the Criminal Investigation Department (CID), headed by Detective Inspector Edmund Reid. Later, Detective Inspectors Frederick Abberline, Henry Moore, and Walter Andrews were sent from Central Office at Scotland Yard to assist. Initially, butchers, surgeons and physicians were suspected because of the manner of the mutilations. The alibi of local butchers and slaughterers were investigated, with the result that they were eliminated from the inquiry. Some contemporary figures thought the pattern of the murders indicated that the culprit was a butcher or cattle drover on one of the cattle boats that plied between London and mainland Europe. Whitechapel was close to the London Docks, and usually such boats docked on Thursday or Friday and departed on Saturday or Sunday. The cattle boats were examined, but the dates of the murders did not coincide with a single boat's movements, and the transfer of a crewman between boats was also ruled out.

At the end of October, Robert Anderson asked police surgeon Thomas Bond to give his opinion on the extent of the murderer's surgical skill and knowledge. The opinion offered by Bond on the character of the "Whitechapel murderer" is the earliest surviving offender profile. Bond's assessment was based on his own examination of the most extensively mutilated victim and the post mortem notes from the four previous canonical murders. In his opinion the killer must have been a man of solitary habits, subject to "periodical attacks of homicidal and erotic mania", with the character of the mutilations possibly indicating "satyriasis". Bond also stated that "the homicidal impulse may have developed from a revengeful or brooding condition of the mind, or that religious mania may have been the original disease but I do not think either hypothesis is likely".

*Handbook for Coroners, police officials, military policemen* was written by the Austrian criminal jurist Hans Gross in 1893, and is generally acknowledged as the birth of the field of criminalistics. The work combined in one system fields of knowledge that had not been previously integrated, such as psychology and physical science, and which could be successfully used against crime. Gross adapted some fields to the needs of criminal investigation, such as crime scene photography. He went on to found the Institute of Criminalistics in 1912, as part of the University of Graz' Law School. This Institute was followed by many similar institutes all over the world. In 1909, Archibald Reiss founded the *Institut de police scientifique* of the University of Lausanne (UNIL), the first school of forensic science in the world. Dr. Edmond Locard, became known as the "Sherlock Holmes of France". He formulated the basic principle of forensic science: "Every contact leaves a trace", which became known as Locard's exchange principle. In 1910, he founded what may have been the first criminal laboratory in the world, after persuading the Police Department of Lyon (France) to give him two attic rooms and two assistants.

Symbolic of the new found prestige of forensics and the use of reasoning in detective work was the popularity of the fictional characterSherlock Holmes, written by Arthur Conan Doyle in the late 19th century. He remains a great inspiration for forensic science, especially for the way his acute study of a crime scene yielded small clues as to the precise sequence of events. He made great use of trace evidence such as shoe and tire impressions, as well as fingerprints, ballistics and handwriting analysis, now known as questioned document examination. Such evidence is used to test theories conceived by the police, for example, or by the investigator himself. All of the techniques advocated by Holmes later became reality, but were generally in their infancy at the time Conan Doyle was writing. In many of his reported cases, Holmes frequently complains of the way the crime scene has been contaminated by others, especially by the police, emphasising the critical importance of maintaining its integrity, a now well-known feature of crime scene examination. He usedanalytical chemistry for blood residue analysis as well as toxicology examination and determination for poisons. He used ballistics by measuring bullet calibres and matching them with a suspected murder weapon.

## *Late 19th-Early 20th Centuries Figures*

Hans Gross applied scientific methods to crime scenes and was responsible for the birth of criminalistics.

Edmond Locard expanded on Gross' work with Locard's Exchange Principle which stated "whenever two objects come into contact with one another, materials are exchanged between them". This means that every contact by a criminal leaves a trace. Locard was also known as the "Sherlock Holmes of France".

Alexander Lacassagne, who taught Locard, produced autopsy standards on actual forensic cases.

Alphonse Bertillon was a French criminologist and founder of Anthropometry (scientific study of measurements and proportions of the human body). He used anthropometry for identification, saying each individual is unique and by measuring aspect of physical difference, there could be a personal identification system. He created the Bertillon System around 1879, which was a way to identify criminals and citizens by measuring 20 parts of the body. In 1884, there was over 240 repeat offenders caught through the Bertillon system. Fingerprinting became more reliable than the Bertillon system.

## *20th century*

Later in the 20th century several British pathologists, Mikey Rochman, Francis Camps, Sydney Smith and Keith Simpson pioneered new forensic science methods. Alec Jeffreys pioneered the use of DNA profiling in forensic science in 1984. He realized the scope of DNA fingerprinting, which uses variations in the genetic code to identify individuals. The method has since become important in forensic science to assist police detective work, and it has also proved useful in resolving paternity and immigration disputes. DNA fingerprinting was first used as a police forensic test to identify the rapist and killer of two teenagers, Lynda Mann and Dawn Ashworth, who were both murdered in Narborough, Leicestershire, in 1983 and 1986 respectively. Colin Pitchfork was identified and convicted of murder after samples taken from him matched semen samples taken from the two dead girls.

Forensic science has been fostered by a number of national forensic science learned bodies including the American Academy of Forensic Sciences (founded 1948), publishers of the *Journal of Forensic Sciences*; the Canadian Society of Forensic Science(founded 1953), publishers of the *Journal of the Canadian Society of Forensic Science*; the British Academy of Forensic Sciences(founded 1960), publishers of *Medicine, science and the law*, and the Australian Academy of Forensic Sciences (founded 1967), publishers of the *Australian Journal of Forensic Sciences*.

## Subdivisions

- Forensic Investigation also known as forensic audit is the examination of documents and the interviewing of people to extract evidence. Forensic investigation is fast emerging as a lucrative professional practice field. With increased sophistication of white collar criminals, there is demand for well trained experts to carry out investigations and also institute preventive, deterrence and detective measures.
- Art forensics concerns the art authentication cases to help research the work's authenticity. Art authentication methods are used to detect and identify forgery, faking and copying of art works, e.g. paintings.
- Computational forensics concerns the development of algorithms and software to assist forensic examination.
- Criminalistics is the application of various sciences to answer questions relating to examination and comparison of biological evidence, trace evidence, impression evidence (such as fingerprints, footwear impressions, and tire tracks), controlled substances,ballistics, firearm and toolmark examination, and other evidence in criminal investigations. In typical circumstances evidence is processed in a Crime lab.
- Digital forensics is the application of proven scientific methods and techniques in order to recover data from electronic / digital media. Digital Forensic specialists work in the field as well as in the lab.
- Ear print analysis is used as a means of forensic identification intended as an identification tool similar to fingerprinting. An earprint is a two-dimensional reproduction of the parts of the outer ear that have touched

a specific surface (most commonly the helix, antihelix, tragus and antitragus).
- Forensic accounting is the study and interpretation of accounting evidence.
- Forensic aerial photography is the study and interpretation of aerial photographic evidence.
- Forensic anthropology is the application of physical anthropology in a legal setting, usually for the recovery and identification of skeletonized human remains.
- Forensic archaeology is the application of a combination of archaeological techniques and forensic science, typically in law enforcement.
- Forensic astronomy uses methods from astronomy to determine past celestial constellations for forensic purposes.
- Forensic botany is the study of plant life in order to gain information regarding possible crimes.
- Forensic chemistry is the study of detection and identification of illicit drugs, accelerants used in arson cases, explosive and gunshot residue.
- Forensic dactyloscopy is the study of fingerprints.
- Forensic document examination or questioned document examination answers questions about a disputed document using a variety of scientific processes and methods. Many examinations involve a comparison of the questioned document, or components of the document, with a set of known standards. The most common type of examination involves handwriting, whereby the examiner tries to address concerns about potential authorship.
- Forensic DNA analysis takes advantage of the uniqueness of an individual's DNA to answer forensic questions such as paternity/maternity testing and placing a suspect at a crime scene, e.g. in a rape investigation.
- Forensic engineering is the scientific examination and analysis of structures and products relating to their failure or cause of damage.
- Forensic entomology deals with the examination of insects in, on and around human remains to assist in determination of time or location of death. It is also possible to determine if the body was moved after death using entomology.
- Forensic geology deals with trace evidence in the form of soils, minerals and petroleum.
- Forensic geomorphology is the study of the ground surface to look for potential location(s) of buried object(s).

- Forensic geophysics is the application of geophysical techniques such as radar for detecting objects hidden underground or underwater.
- Forensic intelligence process starts with the collection of data and ends with the integration of results within into the analysis of crimes under investigation.
- Forensic Interviews are conducted using the science of professionally using expertise to conduct a variety of investigative interviews with victims, witnesses, suspects or other sources to determine the facts regarding suspicions, allegations or specific incidents in either public or private sector settings.
- Forensic limnology is the analysis of evidence collected from crime scenes in or around fresh-water sources. Examination of biological organisms, in particular diatoms, can be useful in connecting suspects with victims.
- Forensic linguistics deals with issues in the legal system that requires linguistic expertise.
- Forensic meteorology is a site-specific analysis of past weather conditions for a point of loss.
- Forensic odontology is the study of the uniqueness of dentition, better known as the study of teeth.
- Forensic optometry is the study of glasses and other eyewear relating to crime scenes and criminal investigations.
- Forensic pathology is a field in which the principles of medicine and pathology are applied to determine a cause of death or injury in the context of a legal inquiry.
- Forensic podiatry is an application of the study of feet footprint or footwear and their traces to analyze scene of crime and to establish personal identity in forensic examinations.
- Forensic psychiatry is a specialized branch of psychiatry as applied to and based on scientific criminology.
- Forensic psychology is the study of the mind of an individual, using forensic methods. Usually it determines the circumstances behind a criminal's behavior.
- Forensic seismology is the study of techniques to distinguish the seismic signals generated by underground nuclear explosions from those generated by earthquakes.
- Forensic serology is the study of the body fluids.
- Forensic social work is the specialist study of social work theories and their applications to a clinical, criminal justice or psychiatric setting. Practitioners of forensic social work connected with the criminal justice

system are often termed Social Supervisors, whilst the remaining use the interchangeable titles Forensic Social Worker, Approved Mental Health Professional or Forensic Practitioner and they conduct specialist assessments of risk, care planning and act as an officer of the court.
- Forensic toxicology is the study of the effect of drugs and poisons on/in the human body.
- Forensic video analysis is the scientific examination, comparison and evaluation of video in legal matters.
- Mobile device forensics is the scientific examination and evaluation of evidence found in mobile phones, e.g. Call History and Deleted SMS, and includes SIM Card Forensics.
- Trace evidence analysis is the analysis and comparison of trace evidence including glass, paint, fibres and hair (e.g., using microspectrophotometry).
- Wildlife Forensic Science applies a range of scientific disciplines to legal cases involving non-human biological evidence, to solve crimes such as poaching, animal abuse, and trade in endangered species.
- Blood Spatter Analysis is the scientific examination of blood spatter patterns found at a crime scene to reconstruct the events of the crime.

## MITIGATING CIRCUMSTANCES

Some countries allow conditions that "affect the balance of the mind" to be regarded as mitigating circumstances. This means that a person may be found guilty of "manslaughter" on the basis of "diminished responsibility" rather than murder, if it can be proved that the killer was suffering from a condition that affected their judgment at the time. Depression, post-traumatic stress disorder and medication side-effects are examples of conditions that may be taken into account when assessing responsibility.

### Insanity

Mental disorder may apply to a wide range of disorders including psychosis caused by schizophrenia and dementia, and excuse the person from the need to undergo the stress of a trial as to liability. In some jurisdictions, following the pre-trial hearing to determine the extent of the disorder, the defence of "not guilty by reason of insanity" may be used to get a not guilty verdict. This defence has two elements:
1. That the defendant had a serious mental illness, disease, or defect.
2. That the defendant's mental condition, at the time of the killing, rendered the perpetrator unable to determine right from wrong, or that what he or she was doing was wrong.

*Under New York law, for example:*

§ 40.15 Mental disease or defect. In any prosecution for an offense, it is an affirmative defence that when the defendant engaged in the proscribed conduct, he lacked criminal responsibility by reason of mental disease or defect. Such lack of criminal responsibility means that at the time of such conduct, as a result of mental disease or defect, he lacked substantial capacity to know or appreciate either: 1. The nature and consequences of such conduct; or 2. That such conduct was wrong.

*Under the French Penal Code:*
- A person is not criminally liable who, when the act was committed, was suffering from a psychological or neuropsychological disorder which destroyed his discernment or his ability to control his actions.
- A person who, at the time he acted, was suffering from a psychological or neuropsychological disorder which reduced his discernment or impeded his ability to control his actions, remains punishable; however, the court shall take this into account when it decides the penalty and determines its regime.

Those who successfully argue a defence based on a mental disorder are usually referred to mandatory clinical treatment until they are certified safe to be released back into the community, rather than prison.

## Post-Partum Depression

Some countries, such as Canada, Italy, Norway, Sweden, the United Kingdom, New Zealand and Australia, allow postpartum depression (also known as post-natal depression) as a defence against murder of a child by a mother, provided that a child is less than two years old (this may be the specific offense of infanticide rather than murder and include the effects of lactation and other aspects of post-natal care). In 2009, Texas state representative Jessica Farrar proposed similar rules for her home state.

## Self-Defence

Acting in self-defence or in defence of another person is generally accepted as legal justification for killing a person in situations that would otherwise have been murder. However, a self-defence killing might be considered manslaughter if the killer established control of the situation before the killing took place. In the case of self-defence it is called a justifiable homicide.

## Unintentional

For a killing to be considered murder, there normally needs to be an element of intent. For this argument to be successful the killer generally needs to demonstrate that they took precautions not to kill and that the death could not have been anticipated or was unavoidable, whatever action they took. As a general rule, manslaughter constitutes reckless killing, while criminally negligent homicide is a grossly negligent killing.

## Diminished Capacity

In those jurisdictions using the Uniform Penal Code, such as California, diminished capacity may be a defence. For example, Dan White used this defence to obtain a manslaughter conviction, instead of murder, in the assassination of Mayor George Moscone and Supervisor Harvey Milk.

## FORENSIC EVIDENCE CASE VIGNETTE

The case of serial child murderer Richard Mark Evonitz highlights the variety of forensic testing that may be utilized to solve difficult cases.

In 1996 and 1997, in Spotsylvania County, Virginia, three young girls were abducted from their residences, sexually assaulted, and killed. The first case occurred on September 9, 1996, when Sophia Silva disappeared from the front porch of her house. She was found in October of 1996, in a swamp, 16 miles from her residence. A suspect was arrested and charged for her murder, based on a faulty trace evidence examination conducted by a state laboratory.

On May 1, 1997, two sisters, Kristin and Kati Lisk, disappeared from their residence after returning home from school. Their bodies were discovered five days later in a river, 40 miles from their residence. After an examination by an FBI Laboratory Examiner yielded trace evidence that positively linked the Silva and Lisk homicides to a common environment, the suspect arrested in the Silva case was subsequently released.

The investigation continued for an additional five years, until a girl was abducted in South Carolina. The victim was able to escape, and she identified Richard Mark Evonitz as her attacker. Evonitz fled South Carolina and was sighted in Florida. After a high-speed chase with police, Evonitz committed suicide. The investigation revealed that Evonitz had lived in Spotsylvania, in 1996 and 1997.

Forensic searches were conducted on Evonitz's residence in South Carolina, his former residence in Spotsylvania, Virginia, and his car. A detailed trace examination of the evidence from these searches and the evidence obtained from the three victims revealed a number of hair and fiber matches, providing sufficient evidence to tie Evonitz to the three murders.

The following trace examinations linked Evonitz to all three homicide victims:
- Fibers from a bath mat.
- Fibers from an afghan.

- Fibers from two separate carpets in Evonitz's former home in Virginia.
- Carpet fibers from the trunk of Evonitz's car.
- Head hair consistent with Evonitz.

A trace examination also linked fibers from a pair of fur-lined handcuffs to the three homicide victims and the surviving victim.

The unique combination of different hair and fiber evidence yielded the "common environment" to which all of the victims and the offender were exposed.

Latent fingerprints belonging to Kristin Lisk were located on the inside of the trunk lid of Evonitz's car, five years after the fact.

## Prosecution of Serial Murder Cases

The recognition and investigation of a serial murder series is often perceived as a separate and distinct process from the other primary goal in these complex cases: the prosecution and conviction of the offender(s) responsible for the homicides. It was a consensus of Symposium attends that law enforcement and prosecutors should work cooperatively as the investigative and prosecution processes are inextricably linked. When police suspect that one or more homicides may be the result of a serial killer, involving the prosecutor early on in the investigation may alleviate significant problems during trial.

The experience of the Symposium attends was that in successful prosecutions of serial murder cases, the prosecutor's office was involved and remained accessible to law enforcement throughout the entire investigation and subsequent arrest. The partnership continued during the trial and resulted in the successful prosecution of the serial murderer. The prosecutor can assist with critical decisions early in the investigation that could potentially impact on court admissibility. Maintaining the integrity of the legal process is a paramount consideration when dealing with court orders, search warrants, Grand Jury testimony, subpoenas, evidence custody matters, capital murder issues, and concerns related to the possible offender's competency and the voluntariness of confessions.

Prosecutors are also in the best position to evaluate the different murder cases within the serial investigation for presentation in court.

They can provide important recommendations regarding the future use of evidence, forensic laboratory work, witness reports, and suspect interviews during trial.

Case management and investigative decision making are still controlled and managed by the law enforcement agencies. The prosecutor acts in an advisory capacity. The responsibilities and duties of the prosecutor should be

## Forensic Sciences

clarified initially in the investigation to avoid potential confusion while the investigation progresses.

In multi-jurisdictional cases, variations in evidentiary standards, search warrant requirements, interview protocols, the quality of the evidence, and the ability to prosecute for capital murder may dictate the appropriate venue for prosecution. This consideration may take on greater significance when the crimes occur in different states.

Expert witnesses often play a significant role in high profile serial murder investigations, dealing with forensic and competency issues. In many investigations and prosecutions, the task of linking the defendant to the victim and the homicide scene(s) has been simplified because of physical, trace, and/or DNA evidence located at the scene. Expert forensic witnesses are utilized to explain the analysis and value of such evidence. Identifying and securing the services of forensic psychologists and psychiatrists will be important when addressing issues of competency, diminished capacity, and the insanity defense. Consideration should also be given for other collateral expert witnesses, who may be utilized to address issues outside of the customary topics, such as blood spatter.

### Prosecution Case Vignette

The Washington, D.C. Beltway sniper attacks serve as an excellent example of multi-jurisdictional prosecutorial considerations. The Beltway serial sniper attacks took place during three weeks of October 2002, in the Washington, D.C. metropolitian area. Ten people were killed and three others critically injured, in various locations throughout the metropolitan area. The killings actually began the month prior to the D.C. rampage, with these offenders committing a number of murders and robberies in several other states. The D.C. area shootings began on October 2nd, with a series of five, fatal shootings over a fifteen-hour period in Montgomery County, Maryland, a suburban county north of Washington, D.C. The investigation was initially spearheaded from Montgomery County, and as the number of shootings mutiplied, the task force involved numerous local, state, and federal law enforcement agencies from Maryland, Virginia, and the District of Columbia. The two men responsible for the homicides, John Allen Muhammad and Lee Boyd Malvo, were eventually captured at an interstate rest area in Western Maryland.

It was ultimately decided that Fairfax County, Virginia, would have the first opportunity to try one of the murders, despite the fact that Maryland had more cases. It was felt that the case in Fairfax was the strongest case. The Fairfax County homicide was the ninth in the Washington, D.C. area series and the third homicide in Virginia. A conviction for murder was secured in this case against Malvo, resulting in a life sentence.

Muhammad was tried next for Capital Murder in a case that occurred in Prince William County, Virginia, which resulted in a death sentence. Malvo, pursuant to a plea agreement, then plead guilty to one count of murder and one attempted murder in Spotsylvania County, Virginia, and was sentenced to life without parole.

Prosecutors in Montgomery County, Maryland, subsequently tried and convicted Muhammad on six counts of murder, and he was sentenced to six consecutive life sentences, without the possibility of parole. Malvo plead guilty and testified against Muhammad. During these trials, Malvo confessed to four other shootings in California, Florida, Texas, and Louisiana. It is unknown whether these or several other jurisdictions, including Arizona, Georgia, Alabama, and Washington State plan to prosecute Muhammad and Malvo.

# 4

# Criminal Psychology and "Fear of Crime"

The fear of crime refers to the fear of being a victim of crime as opposed to the actual probability of being a victim of crime. Studies of the fear of crime occur in criminology. The fear of crime, along with fear of the streets and the fear of youth, is said to have been in Western culture for "time immemorial". Fear of crime can be differentiated into public feelings, thoughts and behaviours about the personal risk of criminal victimization.

These feelings, thoughts and behaviours have a number of damaging effects on individual and group life: they can erode public health and psychological well-being; they can alter routine activities and habits; they can contribute to some places turning into 'no-go' areas via a withdrawal from community; and they can drain community cohesion, trust and neighbourhood stability. Factors influencing the fear of crime include public perceptions of neighbourhood stability and breakdown, circulating representations of the risk of victimization (chiefly via interpersonal communication and the mass media), and broader factors where anxieties about crime express anxieties about the pace and direction of social change. There may also be some wider cultural influences: some have argued that modern times have left people especially sensitive to issues of safety and insecurity.

## Cognitive Aspects of Fear of Crime

Concern about crime can be differentiated from perceptions of the risk of personal victimization (i.e. cognitive aspects of fear of crime). Concern about crime includes public assessments of the size of the crime problem. An example of a question that could be asked is whether crime has increased, decreased or stayed the same in a certain period (and/or in a certain area, for instance the respondents own neighbourhood). Between 1972 and 2001, the Gallup Poll, show that American respondents think crime has decreased. By contrast, the cognitive side of fear of crime includes public perceptions of the likelihood of falling victim, public senses of control over the possibility, and public estimations of the seriousness of the consequences of crime. People who feel especially vulnerable to victimization are likely to feel that they are especially likely to be targeted by criminals (i.e. victimization is likely), that they are unable to control the possibility (i.e. they have low self-efficacy), and that the consequences would be especially severe. Additionally, these three different components of risk perception may interact: the impact of perceived likelihood on subsequent emotional response (worry, fear, anxiety, etc.) is likely to be especially strong among those who feel that consequences are high and self-efficacy is low.

## Behavioural Aspects of Fear of Crime

A third way to measure fear of crime is to ask people whether they ever avoid certain areas, protect certain objects or take preventive measures. This way, measuring fear of crime can become a relatively straightforward thing, because the questions asked tap into actual behaviour and 'objective' facts, such as the amount of money spent on a burglar-alarm or extra locks. However, it is important to note that some degree of 'fear' might be healthy for some people, creating a 'natural defence' against crime. In short, when the risk of crime is real, a specific level of 'fear' might actually be 'functional': worry about crime might stimulate precaution which then makes people feel safer and lowers their risk of crime. The fear of crime is a very important feature in criminology

## Affective Aspects of Fear of Crime

The core aspect of fear of crime is the range of emotions that is provoked in citizens by the possibility of victimization. While people may feel angry and outraged about the extent and prospect of crime, surveys typically ask people "who they are afraid of" and "how worried they are". Underlying the answers that people give are (more often than not) two dimensions of 'fear':
(a) those everyday moments of worry that transpire when one feels personally threatened; and
(b) some more diffuse or 'ambient' anxiety about risk.

While standard measures of worry about crime regularly show between 30% and 50% of the population of England and Wales express some kind of worry about falling victim, probing reveals that few individuals actually worry for their own safety on an everyday basis. One thus can distinguish between fear (an emotion, a feeling of alarm or dread caused by an awareness or expectation of danger) and some more broader anxiety. However it should be noted that some people may be more willing to admit to their worries and vulnerabilities than others.

## Influence of Public Perceptions

Perhaps the biggest influence on fear of crime is public concern about neighbourhood disorder, social cohesion and collective efficacy. The incidence and risk of crime has become linked with perceived problems of social stability, moral consensus, and the collective informal control processes that underpin the social order of a neighbourhood. Such 'day-to-day' issues ('young people hanging around', 'poor community spirit', 'low levels of trust and cohesion') produce information about risk and generate a sense of unease, insecurity and distrust in the environment (incivilities signal a lack of conventional courtesies and low-level social order in public places). Moreover, many people express through their fear of crime some broader concerns about neighbourhood breakdown, the loss of moral authority, and the crumbling of civility and social capital.

People can come to different conclusions about the same social and physical environment: two individuals who live next door to each other and share the same neighbourhood can view local disorder quite differently. Why might people have different levels of tolerance or sensitivity to these potentially ambiguous cues? UK research has suggested that broader social anxieties about the pace and direction of social change may shift levels of tolerance to ambiguous stimuli in the environment. Individuals who hold more authoritarian views about law and order, and who are especially concerned about a long-term deterioration of community, may be more likely to perceive disorder in their environment (net of the actual conditions of that environment). They may also be more likely to link these physical cues to problems of social cohesion and consensus, of declining quality of social bonds and informal social control.

## Police as Social Institution

The rise of institutions always depends on common goals, which can be very different and basically depend on recognizing the same, or similar, needs and interests. At the beginning, recognizing common problems is sufficient for the definition of, and agreement on, joint goals. The permanence of some common necessities requires a long duration of similar actions, but for successful and lasting operations it becomes more necessary than recognizing common problems. It is more than enough to reinstate a stable arrangement of common rules or norms, which is possible in two main ways, although they never appear in a pure form.

The first is habitualization. The process of habitualization supports stabilization of mutual relationships between elements of an institution. "The processes of habitualization precedes any institutionalization... Institutionalization occurs whenever there is a reciprocal typification of habitualized actions by types of actors. Put differently any such typification is an institution. What must be stressed is the reciprocity of institutional typifications and the typicality of not only the actions but also the actors in institutions. The typifications of habitualized actions that constitute institutions are always shared ones. They are available to all the members of the particular social group in question, and the institution itself typifies individual actors as well as individual actions in stitutions further imply historicity and control.

Reciprocal typifications of actions are built up in the course of a shared history. Institutions always have a history of which they are the product. It is impossible to understand an institution adequately without an understanding of the historical process in which it was produced. They also by the very fact of their existence, control human conduct by setting up predefined patterns of conduct, which channel it in one direction as against the many other directions that would theoretically be possible. In actual experience institutions generally manifest themselves in collectivities containing considerable numbers of people". The second is social contract, a concept developed by Thomas Hobbes. A social contract is a solution when, once one has reached a contractual agreement with another, he grants his approval to the obligations and the correlative rights that accrue to him and to his partner on the basis of this agreement.

On the level of social structure, individuals (status and roles) and groups who try to achieve similar goals make efforts to find, for all of them, acceptable ways of mutual communication, and ways of conducting mutual activities, because they make a contract about the basis of their work: mutual relationships; useful instruments, and hierarchy of goals. With time, arranged norms become obligatory in any individual case, and in accordance with this, become indisputably the basis of any activities. Habitualization and social contract are two essential conditions in the process of stabilizing institutions. At same time the process of legitimation of institution toward outside social reality continues: "The function of legitimation is to make objectively available and subjectively plausible the 'first order' objectivations that have been institutionalized. While we define legitimation by this function, regardless of the specific motives inspiring any particular legitimating process, it should be added that 'integration' in one form or another, is also the typical purpose motivating the legitimators".

"Legitimatization is not necessary in the first phase of institutionalization, when the institution is simply an act that requires no further support either intersubjectively or biographically; it is self-evident to all concerned. The problem of legitimation inevitably arises when the objectivations of the new historic institutional order are to be transmitted to a new genaration". John

Rawls (1999.) emphasizes justice as the first virtue of social institutions, as truth is of systems of thought. Each person possesses an inviolability founded on justice that even the welfare of society as a whole cannot override. Society is well ordered when it is not only designed to advance the good of its members but when it is also effectively regulated by a public conception of justice. That is, it is a society in which:
1. everyone accepts and knows that the others accept the same principles of justice, and
2. the basic social institutions generally satisfy and are generally known to satisfy these principles.

He also explains the well-ordered peoples' right to war: "No state has a right to war in the pursuit of its rational, as opposed to its reasonable, interests. The Law of Peoples does, however, assign to all well-ordered peoples (both liberal and docent), and indeed to any society that follows and honors a reasonably just Law of Peoples, the right to war in self-defence. When a liberal society engages in war in selfdefense, it does so to protect and preserve the basic freedoms of its citizens and its constitutionally democratic political institutions". The role of social institutions is to meet social needs. Sociologists define the role in terms of expectations: groups of norms and values that individuals and institutions must be able to achieve.

- This world and its citizens need security: a life without war, without terorrism, without crime, without poverty, and without threat of contaminated air, water, earth, flora and fauna;
- such a world may be built only with respect for the phenomena of life (this has the highest value because it is irretrievable), mutual respect, dignity, cooperation, a sense of justice and solidarity (economic, educational, cultural; other institutions also have the role of caring about human needs);
- on the other hand, there always exists a need for more selfish success, more power, more control and more manipulation of people and their property (the military and police are institutions whose role is to prevent this).

Max Weber (1978.) studied the role of military services throughout human history and compiled a list of services that this type of institution performs in the life of society. "Discipline as the basis of warfare, gave birth to patriarchal kingship among the Zulus, where the monarch, however, was constitutionally limited by the power of the army commanders – similar to the (manner in which) the) Spartan (kings were checked by the) ephors. Similarly, discipline gave birth to the Hellenic polis with gymnasia. When infantry drill was perfected to the point of virtuosity (as in Sparta), the polis had inevitably an aristocratic structure; when cities resorted to naval discipline, they had a democratic structure (Athens). Military discipline was also the basis of Swiss democracy, which in heyday of the Swiss mercenaries was very different from the Athenian but controlled – in Greek terms – territories with inhabitants of limited rights (perioeci) or with no rights (helots). Military discipline was also instrumental in establishing the rule of the Roman patriciate and, finally, the bureaucratic states of Egypt, Assyria and modern Europe..

The well-trained Spartan Army, the organization of the other Hellenic and Macedonian and of several Oriental military establishments, the Turkish quasi– prebendal fiefs, and finally the feudal fiefs of the Japanese and Occidental Middle Ages – all of these were stages of the economic decentralization which usually goes hand in hand with the weakening of discipline and the rise of individual heroism. From the disciplinary aspect, just as from the economic, the seigneurial vassal represents an extreme contrast to the patronomial or bureaucratic soldier. And the disciplinary aspect, is a consequence of the economic aspect… Military discipline gives birth to all discipline". In this capacity the institutionalization of military and police serves society. Always, in any past or future cases, the first and main role of these types of institutions is providing safety to other people. Problems exist and will probably always exist about different interpretations regarding the instruments this role includes, and what circumstances must exist for complete moral justification of the use of force in protecting people without arms and without bad intentions, against individuals and groups with arms and with bad intentions.

The role of military and police services (institutions) in the global, civilian and democratic world of the future We now turn to the ways in which the military and police can constructively co-exist in a civilian, democratic and global human society: offering mutual support, a clear strategy of action, and behaving responsibly towards citizens. As we have said, social institutions are structured answers to basic needs of human society. They originate, develop, and live as long they present the best way to solving real social problems, or as long those problems exist in social environment. Some needs, such as individual and social need for safety, have never changed, and so we are witnesses of continually developing ways in which human societies respond to the many threats to human existence. During the history of civilization, the military and police institutions have changed their shape (their level and forms of organization), but they have existed in every society, regardless of its level of development. After states broke up or regimes changed, the first institutions that arose faster than others were, usually or very often, those two.

The reason for the changes in the military and police institutions has been the loss of orientation of action, but never the purpose of their existence. When they became instruments of conquest instead of defence; instruments of violence instead of keeping the peace; instruments used against the citizens (within a state or outside of it) instead of protecting their rights, stability and safety, they either broke down or changed, sometimes immediately, sometimes after a number of years, but they invariably did. Today, at the beginning of a new millennium, the purpose of their existence is changing: threats are becoming different, the environment is becoming different and, consequently, the military and police must follow the rhythm of those changes. The processes of globalization have transformed the ways of living and ways of thinking. Particular goals, interests and responsibilities, those of particular states or countries, have ceased to exist whether they want it or not, and common goals, interests and responsibilities have taken over.

We are witnesses that the processes of globalization in many cases have their own laws and do not depend on people's wishes. Interestingly, threats spread faster than the opportunities and resources for neutralizing them: global terrorism almost has replaced the conventional warfare among states; the international organized crime (gun-running, drug-and human trafficking) has prevailed instead of the usual types of crime; and environmental destruction has taken the shape of destruction of life on Earth. Sometimes it seems that the "institutions" of the world crime are becoming global faster and easier than the institutions that must stop them. The common interests of neutralizing threats to the present and future require only one goal: building institutions, both military and police, that would be able to protect global safety. It means life without terrorism, without war, without crime, without poverty created by injustice, and without ecological catastrophes caused by a blind wish for more profit.

How can those institutions meet this challenge? It seems logical and obvious that neutralizing the threats to the world's citizens will require institutions that are not limited by state borders, but are able to solve global problems. In these new circumstances, the classic concept of national (state) armed forces would not able to neutralize global threats. The right answer to those problems must also be global. This means that the military and police institutions have to transform themselves as soon as possible, not only to become more civilian and democratic, but also more globalized. Among the contemporary papers focusing on relationships between military services and society, an important study is "The Armed forces and Society" by Timothy Edmunds, Anthony Forster and Andrew Cottey (2002.), who provide a survey of recent literature on this subject, citing Martin Edmonds, James Gow, Christopher Dandeker, Charles Moskos, Frank Wood and Martin Shaw.

Edmunds suggests that the nature of the military's task – where personnel are expected to have 'unlimited liability' and the prospect of being killed is almost a 'definitional' aspect of service – does make it different from other institutions in society such as the police or the civil-service. These tasks, he argues, necessitate the transfer of individual values to those of the group, and require the maintenance of particularly high levels of moral and discipline. As a result, Edmonds suggests that the 'armed services fulfil a highly specialized function, the effect of which is to separate them entirely, and geographically to a great extent, from civil society'. Gow takes a different approach, identifying legitimacy as the key element in the relationship between the military and society. Gow argues that military legitimacy has both functional and socio-political bases. The functional basis of military legitimacy derives from its 'military mission' – which he defines as the protection of the state from external threat. The socio-political bases of military legitimacy are more complex, and stem from the nature of the military's relationship with political authority, its role as a symbol of political unity and national pride (informed by military traditions and past military activities), its contribution to the socio-economic infrastructure of the state, and its role as an instrument of education and socialisation.

This military legitimacy is the basis of a 'social contract' between soldiers and the socio-political community, and for Gow, 'support for the military] will be forthcoming if it] performs effectively, in accordance with its functional and socio-political bases of legitimacy, or if there is some attachment to those bases that overarch poor performance'. Christopher Dandeker notes that while the armed forces share 'institutional qualities' – such as the need for teamwork, leadership and loyalty to the organisation – with other civilian organisations, their war-fighting role necessitates a level of coercion in military discipline which sets them apart. Thus, he observes that 'the functional imperative of war ensures that the military will always stand apart from civilian society'. Dandeker also addresses the second military and society debate. He observes that despite the demands of the military's functional imperative, a series of challenges have emerged to military culture and its 'right to be different' which amount to 'new times' for the military. These include, first, a changed strategic context – which in the west entails an end to immediate threats to national security and a more 'globalised' world where challenges to state sovereignty have come from both above from supranational organisations and below from regionalism and a globalisation of social and cultural relationships.

Second, a changed societal context in which the supporting framework for core military values is increasingly challenged by a more individualistic, egalitarian and litigious society. Finally and partly as a result of the changed strategic context, increasing cost pressures which have led to the civilianisation of many traditionally military jobs such as logistical support. Charles Moskos and Frank Wood have suggested that societal pressures are leading western militaries to shift from an institutional structure to a more civilianised organizational one.

Moskos et al argue that these changes are significant enough to be considered a new, postmodern phase of military organisation and military-society relations. For them, the postmodern military is characterised by an increasing inter-penetrability between civilian and military spheres; a diminution of differences within the military itself, particularly between different ranks and services; non-traditional military operations such as peacekeeping; the increasing importance and prevalence of supranational or multinational command structures or at least legitimation for military operations; the internationalisation of the military themselves. Martin Shaw argues his twin conceptions of post-military and common risk society. For Shaw, the Cold War period in much of the industrialised world was characterized by the militarisation of society through the necessity for mass armies and conscriptions. Moreover, in many states such as France, this militarisation was reinforced by a conception of society which emphasised a contract between the state and its citizens. Thus, in return for their rights, citizens were expected to provide service to the state through conscription.

Shaw argues that geopolitical changes coupled with economic growth and a revolution of rising expectations are increasingly undermining this militarisation leading to a post-military society where military service and experience are the exception rather than the norm. Shaw has refined this theme, arguing that the militarisation along national lines so characteristic of the Cold War period has been replaced by a common risk society, in which perceptions of threat from problems such as global warming are perceived to be increasingly transnational. In conjunction with the changes associated with the post-military society, this shift of perception has resulted in the replacement of traditional military symbols and places in national cultures with spectator sport militarism, where societal engagement with the military is limited to passive observance through the media.

The building this type of institution requires deep deliberation about the basic idea of serving common interests, based on unconditional respect for human life and dignity as irretrievable values; on the other hand, it means avoiding inhibited factors as particular, special and, with increasing frequency, states' interests.

On the strategic level of concrete activity, it requires changes in the use of power in social institutions (especially military, police and governments) from the traditional use of power to achieve special (selfish) interests, to the use of power to create the common good for the citizens. There are many examples (from ancient to recent history) that would show that the traditional use of power for achieving particular interests becomes dissipation of energy, and finally makes existing problems even worse. Just several very tragic examples: the indulgence of Adolph Hitler, Mohamed Farrah Adeed, or Slobodan Miloševiæ as a result of some powerful politicians and governments behaving with short-term and short-range interests in mind at the beginning, which created an irrevocable tragedy afterwards.

The conclusion is that tolerating and allowing the violation of basic principles of humanity and democracy always leads to a complete loss of control subsequently, which has proved to be a social and historical fact. The only logical solution is to create institutions that would be able to operate on the same rules and principles to prevent any individual case, anywhere. Some such institutions already exist, but their activities could be improved to support particular states, interest groups and individuals. The role of what could be described as a common government could be built on the idea of the United Nations if all the states concerned invested more of their own instruments of power and sovereignty in such a joint governing body. Those decisions would be possible if they were based on free will and a clear vision about the benefits that would result from such a move. Only then would the UN be able to direct the joint power to improve our co-existence.

NATO is an organization which comes close to this way of thinking and operating. But the necessary level for overcoming the global threats is hindered by the organization's particular and isolated interests. In NATO's case, the lack of vision and the orientation towards particular states' interests is not associated with military professionals, but with political decision makers. For example, in the case of Bosnia and Kosovo in the 1990s, decisions on the type of engagement hinged on political compromises.

Again and again, mixing political interests with expert knowledge creates results which are bad, or at least not good enough. However, NATO has the capacity to become an institution capable of bringing together the skills and knowledge necessary for using joint power to neutralize threats and protect the common good. In the real world today, there are governments whose terms in office are marked by responsible and altruistic, as well as irresponsible, egoistic, or even criminal decisions at the same time. The deal of liberal democracy overcomes this latter way of thinking and behaviour.

There are two fundamental conditions that must be realized at a very high level, which can counteract any undemocratic influence. The first, most frequently mentioned notion is human rights. Blandine Kriegel believes that three conditions are necessary for a doctrine of human rights. First, the human being as such must be recognized as a having value. Second, this recognition must be given legal expression. Finally, this legal status must be guaranteed by political authorities.

The next most important concept is the rule of law among persons and among peoples. John Rawls cites several authors who made lists similar to the principles of international law: peoples are free and independent, and their freedom and independence are to be respected by other peoples; peoples are to observe treaties and undertakings; peoples are equal and are parties to the agreements that bind them; peoples are to observe a duty of non-intervention; peoples have the right of self-defence but no right to instigate war for reasons other than self-defence; peoples are to honor human rights; peoples are to observe certain specified restrictions in the conduct of war; peoples have a duty to assist other peoples living under unfavourable conditions that prevent their having a just or decent political and social regime.

In short, if any decisions, especially strategic ones, are based on the rule of law and show respect for the human rights of all those persons that will bear the consequences of these decisions, there is a strong possibility that the results will be far-reaching and positive. On the contrary, making decisions without respect for human rights and the rule of law will always produce bad consequences, which creates new problems and threats, with their own unpredictable laws.

On what does the quality, duration, and legitimacy of strategic decisions depend? Considering the decision-making process, especially regarding those decisions governing the organized types of social activities, we can identify three kinds of logic, that is, three kinds of interests that go into the creation of strategic decisions. These three kinds of logic are bureaucratic, political and expert, since every level of strategic decision making (within a state, at the state level and at the international level) includes those three groups of people, and those three ways of thinking. Each one of them has a very important role, but problems emerge if any of them pretends to play the key role in the decision-making process. Bureaucratic logic is based on misgivings about any changes, especially those done quickly, and bureaucratic vision is always based on tradition. Usually, tradition is the main reason and has the advantage of critical consideration about usefulness or purposefulness. Political logic is usually guided by political success, which must be evident, quick and useful in increasing the popularity of a particular political party.

This kind of purpose is usually the main reason in the reaching of decisions. Expert logic is based on empirical validity. It includes all established information connected with a certain problem. This logic tries to discover the roots, because they are the best answers about the nature of the problem. Serious consideration about all possible predictable consequences is a very important part of expert decision making.

The problem with this way of thinking is usually connected with yielding in duels with political and administrational authorities on the one hand, and, on the other hand, with their feeling of superiority, which often causes them to ignore some very serious common problems. The best (ideal) decisions start when the persons chosen by citizens in elections – the politicians – recognize real social needs and problems, and prioritize them with the help of experts (depending on the kind of threat and the predictable bad consequences that can arise if the threat is ignored). Experts then suggest solutions, means, and all possible ways for solving the problem, and list all predictable consequences. The bureaucratic segment takes care of supporting them by background (literature, similar cases in tradition), public feedback, and whether the measures considered usually work, most importantly regarding the normal administration and state life.

At the same time, every proposal is checked in light of human rights and the rule of law. The final decision is a result of a concerted effort, and enjoys strong support by non-governmental agencies, which are involved to offer special advice on human rights. If the presentation of the decision and the request for support of the citizens and all interested parties (the international partners, organizations and citizens) includes illustrating the whole decision-making process with all the efforts it involved, universal support is almost guaranteed. Which influence was greater than necessary is usually clear from the consequences. As we have said, responsible decisions are recognized as being long-term, right, legitimate and enjoying the support of the citizens. If those qualities are lacking, Renato Matiæ: The Social Role of Military and Police Institutions... there is not much chance for fixing the situation, because the bad consequences have already created even more serious problems.

This is still the main approach. Although most world leaders pledge to uphold the principles of liberal democracy (human rights and the rule of law), their concrete decisions are still based on a very primitive use of power. The reasons for it are obvious. Depending on power always means and requires an investment in power. Standing by those principles is very expensive. If a lot of money and reputations are invested in a certain decision, or a number of decisions, which then prove to have been erroneous, admitting this fact would be very difficult and risky. A more convenient solution is to bend the facts to suit the current situation.

This means persistently defending the wrong way, with more and more deposits of power and money. Politics that are based on this kind of idea can bring together many clever and educated people in order to explain and exculpate a certain orientation, who would create political theories based on twisted truths. It can even become a dominant way of political thinking and educate many generations of political thinkers. They can even create the public opinion that this is the only real strategy in global politics. It is possible to defend any policy by using force, but this policy, if it is based on irresponsible decisions, always produces injustice, violence and new threats. With time it creates a critical amount of displeasure and animosity, and breaks down, but unfortunately bad consequences remain. They are often irreparable or require a lot of good will and sacrifice to be rectified.

How is it possible to continue doing the wrong thing when all facts that argue to the contrary are so evident? Does anyone strongly believe that the same way of thinking that has caused many of today's and future threats in the first place can now create the right solutions for neutralizing them? "Policymakers are forever tempted to wait for a case to arise before dealing with it; manipulation replaces reflection as the principal policy tool. But the dilemmas of foreign policy are not only – or perhaps even primarily – the by-product of contemporary events; rather they are the end-product of the historical process that shaped them. Modern decision making is overwhelmed not only by contemporary facts but by the immediate echo which overwhelms perspective. Instant punditry and the egalitarian conception that any view is as valid as any other combine with a cascade of immediate symptoms to crush a sense of perspective".

## CRIME ASSESSMENT STAGE

The Crime Assessment Stage in generating a criminal profile involves the reconstruction of the sequence of events and the behaviour of both the offender and victim. Based on the various decisions of the previous stage, this reconstruction of how things happened, how people behaved, and how they planned and organized the encounter provides information about specific characteristics to be generated for the criminal profile.

Assessments are made about the classification of the crime, its organized/disorganized aspects, the offender's selection of a victim, strategies used to control the victim, the sequence of crime, the staging (or not) of the crime, the offender's motivation for the crime, and crime scene dynamics.

The classification of the crime is determined through the decision process outlined in the first decision process model. The classification of a crime as organized or disorganized, first introduced as classification of Lust murder, but since broadly expanded, includes factors such as victim selection, strategies to control the victim, and sequence of the crime.

An organized murderer is one who appears to plan his murders, target his victims, display control at the crime scene, and act out a violent fantasy against the victim (sex, dismemberment, torture). For example, Ted Bundy's planning was noted through his successful abduction of young women from highly visible areas (e.g., beaches, campuses, a ski lodge). He selected victims who were young, attractive, and similar in appearance. His control of the victim was initially through clever manipulation and later physical force. These dynamics were important in the development of a desired fantasy victim.

In contrast, the disorganized murderer is less apt to plan his crime in detail, obtains victims by chance and behaves haphazardly during the crime.

For example, Herbert Mullin of Santa Cruz, California who killed 14 people of varying types (e.g., an elderly man. a young girl, a priest) over a four-month period did not display any specific planning or targeting of victims; rather, the victims were people who happened to cross his path, and their killings were based on psychotic impulses as well as on fantasy.

The determination of whether or not the crime was staged (i.e., if the subject was truly careless or disorganized. or if he made the crime appear that way to distract or mislead the police) helps direct the investigative profiles to the killer's motivation. In one case a 16-year-old high school junior living in a small town failed to return home from school.

Police, responding to the father's report of his missing daughter, began their investigation and located the victim's scattered clothing in a remote area outside the town. A crude map was also found at the scene which seemingly implied a premeditated plan of kidnaping. The police followed the map to a location which indicated a body may have been disposed of in a nearby river. Written and telephoned extortion demands were sent to the father. a bank executive, for the sum of $80,000, indicating that a kidnap was the basis of the abduction. The demands warned police in detail not to use electronic monitoring devices during their investigative efforts.

Was this crime staged? The question was answered in two ways. The details in one aspect of the crime (scattered clothing and tire tracks) indicated that subject was purposely staging a crime while the details in the other (extortion) led the profiles to speculate who the subject was; specifically that he had a law enforcement background and therefore had knowledge of police procedures concerning crimes of kidnaping, hiding the primary intent of sexual assault and possible murder. With this information, the investigative profiles recommended that communication continue between the suspect and the police, with the hypothesis that the behaviour would escalate and the subject become bolder.

While further communications with the family were being monitored, profiles from the FBI's Behavioural Science Unit theorized that the subject of the case was a white male who was single, in his late 20's to early 30's, unemployed, and who had been employed as a law enforcement officer within the past year.

He would be a macho outdoors type person who drove a late model, well maintained vehicle with a CB radio. The car would have the overall appearance of a police vehicle.

As the profile was developed the FBI continued to monitor the extortion telephone calls made to the family by the subject. The investigation, based on the profile, narrowed to two local men, both of whom were former police officers.

One suspect was eliminated, but the FBI became very interested in the other since he fit the general profile previously developed. This individual was placed under surveillance. He turned out to be a single, white male who was previously employed locally as a police officer.

He was now unemployed and drove a car consistent with the FBI profile. He was observed making a call from a telephone booth, and after hanging up, he taped a note under the telephone. The call was traced to the residence of the victim's family. The caller had given instructions for the family to proceed to the phone booth the suspect had been observed in.

"The instructions will be taped there," stated the caller. The body of the victim was actually found a considerable distance from the "staged" crime scene, and the extortion calls were a diversion to intentionally lead the police investigation away from the sexually motivated crime of rape-murder.

The subject never intended to collect the ransom money, but he felt that the diversion would throw the police off and take him from the focus of the rape-murder inquiry. The subject was subsequently arrested and convicted of this crime.

## Motivation

Motivation is a difficult factor to judge because it requires dealing with the inner thoughts and behaviour of the offender. Motivation is more easily determined in the organized offender who premeditates, plans, and has the ability to carry out a plan of action that is logical and complete. On the other hand, the disorganized offender carries out his crimes by motivations that frequently are derived from mental illnesses and accompanying distorted thinking (resulting from delusions and hallucinations). Drugs and alcohol, as well as panic and stress resulting from disruptions during the execution of the crime, are factors which must be considered in the overall assessment of the crime scene.

## Crime Scene Dynamics

Crime scene dynamics are the numerous elements common to every crime scene which must be interpreted by investigating officers and are at times easily misunderstood. Examples include location of crime scene, cause of death, method of killing, positioning of body, excessive trauma, and location of wounds.

The investigative profiles reads the dynamics of a crime scene and interprets them based on his experience with similar cases where the outcome is known. Extensive research by the Behavioural Science Unit at the FBI Academy and indepth interviews with incarcerated felons who have committed such crimes have provided a vast body of knowledge of common threads that link crime scene dynamics to specific criminal personality patterns. For example, a common error of some police investigators is to assess a particularly brutal lust-mutilation murder as the work of a sex fiend and to direct the investigation toward known sex offenders when such crimes are commonly perpetrated by youthful individuals with no criminal record.

## CRIMINAL PROFILE STAGE

The fourth stage in generating a criminal profile deals with the type of person who committed the crime and that individual's behavioural organization with relation to the crime. Once this description is generated, the strategy of investigation can be formulated, as this strategy requires a basic understanding of how an individual will respond to a variety of investigative efforts. Included in the criminal profile are background information (demographics), physical characteristics, habits, beliefs and values, pre-offense behaviour leading to the crime, and post-offense behaviour. It may also include investigative recommendations for interrogating or interviewing, identifying, and apprehending the offender.

This fourth stage has an important means of validating the criminal profile-Feedback No. 1. The profile must fit with the earlier reconstruction of the crime, with the evidence, and with the key decision process models. In addition, the investigative procedure developed from the recommendations must make sense in terms of the expected response patterns of the offender. If there is a lack of congruence, the investigative profiles review all available data. As Hercule Poirot observed, "To know is to have all of the evidence and facts fit into place."

### Investigation Stage

Once the congruence of the criminal profile is determined, a written report is provided to the requesting agency and added to its ongoing investigative efforts. The investigative recommendations generated in Stage 4 are applied, and suspects matching the profile are evaluated. If identification, apprehension, and a confession result, the goal of the profile effort has been met. If new evidence is generated (e.g., by another murder) and/or there is no identification of a suspect, re-evaluation occurs via Feedback No. 2. The information is re-examined and the profile revalidated.

## Apprehension Stage

Once a suspect is apprehended, the agreement between the outcome and the various stages in the profile-generating-process are examined. When an apprehended suspect admits guilt, it is important to conduct a detailed interview to check the total profiling process for validity.

## Case Example

A young woman's nude body was discovered at 3:00 p.m. on the roof landing of the apartment building where she lived. She had been badly beaten about the face and strangled with the strap of her purse. Her nipples had been cut off after death and placed on her chest. Scrawled in ink on the inside of her thigh was, "You can't stop me." The words "Fuck you" were scrawled on her abdomen. A pendant in the form of a Jewish sign (Chai), which she usually wore as a good luck piece around her neck, was missing and presumed taken by the murderer. Her underpants had been pulled over her face; her nylons were removed and very loosely tied around her wrists and ankles near a railing. The murderer had placed symmetrically on either side of the victim's head the pierced earrings-she had been wearing. An umbrella and inkpen had been forced into the vagina and a hair comb was placed in her pubic hair. The woman's jaw and nose had been broken and her molars loosened. She suffered multiple face fractures caused by a blunt force. Cause of death was asphyxia by ligature (pocketbook strap) strangulation. There were post-mortem bite marks on the victim's thighs, as well as contusions, hemorrhages, and lacerations to the body. The killer also defecated on the roof landing and covered it with the victim's clothing.

## Profiling Inputs

In terms of crime scene evidence, everything the offender used at the crime scene belonged to the victim. Even the comb and the felt-tip pen used to write on her body came from her purse. The offender apparently did not plan this crime; he had no gun, ropes, or tape for the victim's mouth. He probably did not even plan to encounter her that morning at that location. The crime scene indicated a spontaneous event; in other words, the killer did not stalk or wait for the victim. The crime scene differs from the death scene. The initial abduction was on the stairwell; then the victim was taken to a more remote area.

Investigation of the victim revealed that the 26-year-old, 90-pound, 4' 11" white female awoke around 6:30 a.m. She dressed, had a breakfast of coffee and juice, and left her apartment for work at a nearby day care center, where she was employed as a group teacher for handicapped children. She resided with her mother and father. When she would leave for work in the morning, she would take the elevator or walk down the stairs, depending on her mood. The victim was a quiet young woman who had a slight curvature of the spine (kyhoscoliosis).

The forensic information in the medical examiner's report was important in determining the extent of the wounds, as well as how the victim was assaulted and whether evidence of sexual assault was present or absent. No semen was noted in the vagina, but semen was found on the body. It appeared that the murderer stood directly over the victim and masturbated. There were visible bite marks on the victim's thighs and knee area. He cut off her nipples with a knife after she was dead and wrote on the body. Cause of death was strangulation, first manual, then ligature, with the strap of her purse. The fact that the murderer used a weapon of opportunity indicates that he did not prepare to commit this crime.

He probably used his fist to render her unconscious, which may be the reason no one heard any screams. There were no deep stab wounds and the knife used to mutilate the victim's breast apparently was not big, probably a penknife that the offender normally carried. The killer used the victim's belts to tie her right arm and right leg, but he apparently untied them in order to position the body before he left.

The preliminary police report revealed that another resident of the apartment building, a white male, aged 15, discovered the victim's wallet in a stairwell between the third and fourth floors at approximately 8:20 a.m. He retained the wallet until he returned home from school for lunch that afternoon. At that time, he gave the wallet to his father, a white male, aged 40. The father went to the victim's apartment at 2:50 p.m. and gave the wallet to the victim's mother.

When the mother called the day care center to inform her daughter about the wallet, she learned that her daughter had not appeared for work that morning. The mother, the victim's sister, and a neighbour began a search of the building and discovered the body. The neighbour called the police. Police at the scene found no witnesses who saw the victim after she left her apartment that morning.

## Decision Process

This crime's style is a single homicide with the murderer's primary intent making it a sexually motivated type of crime. There was a degree of planning indicated by the organization and sophistication of the crime scene. The idea of murder had probably occupied the killer for a long period of time. The sexual fantasies may have started through the use and collecting of sadistic pornography depicting torture and violent sexual acts.

Victim risk assessment revealed that the victim was known to be very self-conscious about her physical handicap and size and she was a plain-looking woman who did not date. She led a reclusive life and was not the type of victim that would or could fight an assailant or scream and yell. She would be easily dominated and controlled, particularly in view of her small stature.

Based upon the information on occupation and lifestyle, we have a low-risk victim living in an area that was at low risk for violent crimes. The apartment building was part of a 23-building public housing project in which the racial mixture of residents was 50% black, 40% white, and 10% Hispanic. It was located in the confines of a major police precinct. There had been no other similar crimes reported in the victim's or nearby complexes.

The crime was considered very high risk for the offender. He committed the crime in broad daylight, and there was a possibility that other people who were up early might see him. There was no set pattern of the victim taking the stairway or the elevator. It appeared that the victim happened to cross the path of the offender.

There was no escalation factor present in this crime scene. The time for the crime was considerable. The amount of time the murderer spent with his victim increased his risk of being apprehended. All his activities with the victim-removing her earrings, cutting off her nipples, masturbating over her-took a substantial amount of time.

The location of the crime suggested that the offender felt comfortable in the area. He had been here before, and he felt that no one would interrupt the murder.

## Crime Assessment

The crime scene indicated the murder was one event, not one of a series of events. It also appeared to be a first-time killing, and the subject was not a typical organized offender. There were elements of both disorganization and organization; the offender might fall into a mixed category. A reconstruction of the crime/death scene provides an overall picture of the crime. To begin with the victim was not necessarily stalked but instead confronted. What was her reaction? Did she recognize her assailant, fight him off, or try to get away'? The subject had to kill her to carry out his sexually violent fantasies. The murderer was on known territory and thus had a reason to be there at 6:30 in the morning: either he resided there or he was employed at this particular complex.

The killer's control of the victim was through the use of blunt force trauma, with the blow to her face the first indication of his intention. It is probable the victim was selected because she posed little or no threat to the offender. Because she didn't fight, run, or scream, it appears that she did not perceive her abductor as a threat. Either she knew him, had seen him before, or he looked non-threatening (i.e., he was dressed as a janitor, a postman, or businessman) and therefore his presence in the apartment would not alarm his victim.

In the sequence of the crime, the killer first rendered the victim unconscious and possibly dead; he could easily pick her up because of her small size. He took her up to the roof landing and had time to manipulate her body while she was unconscious. He positioned the body, undressed her, acted out certain fantasies which led to masturbation.

The killer took his time at the scene, and he probably knew that no one would come to the roof and disturb him in the early morning since he was familiar with the area and had been there many times in the past. The crime scene was not staged. Sadistic ritualistic fantasy generated the sexual motivation for murder. The murderer displayed total domination of the victim. In addition, he placed the victim in a degrading posture, which reflected his lack of remorse about the killing.

The crime scene dynamics of the covering of the killer's feces and his positioning of the body are incongruent and need to be interpreted.

First, as previously described, the crime was opportunistic. The crime scene portrayed the intricacies of a long-standing murderous fantasy. Once the killer had a victim, he had a set plan about killing and abusing the body. However, within the context of the crime, the profiles note a paradox: the covered feces.

Defecation was not part of the ritual fantasy and thus it was covered. The presence of the feces also supports the length of time taken for the crime, the control the murderer had over the victim (her unconscious state), and the knowledge he would not be interrupted. The positioning of the victim suggested the offender was acting out something he had seen before, perhaps in a fantasy or in a sado-masochistic pornographic magazine. Because the victim was unconscious, the killer did not need to tie her hands. Yet he continued to tie her neck and strangle her. He positioned her earrings in a ritualistic manner, and he wrote on her body. This reflects some son of imagery that he probably had repeated over and over in his mind. He took her necklace as a souvenir; perhaps to carry around in his pocket. The investigative profiles noted that the body was positioned in the form of the woman's missing Jewish symbol.

## CRIMINAL PSYCHOLOGIST

A criminal psychologist is a professional that studies the behaviours and thoughts of criminals. Thanks to a number of popular television programs such as *Criminal Minds* that depict fictionalized criminal psychologists, interest in this career field has grown dramatically in recent years. The field is highly related to forensic psychology and in some cases the two terms are used interchangeably. What is a career in criminal psychology really like? Is it as exciting as it looks in all those tv dramas? Continue reading to learn more about criminal psychologists, including exactly what they do, where they work and the type of education and training it takes to enter this profession.

What Does a Criminal Psychologist Do?: A large part of what a criminal psychologist does is study why people commit crimes. However, they may also be asked to assess criminals in order to evaluate the risk of recidivism or make educated guesses about the actions that a criminal may have taken after committing a crime.

In addition to helping law enforcement solve crimes or analyse the behaviour of criminal offenders, criminal psychologists are also often asked to provide expert testimony in court. Perhaps one of the best known duties of a criminal psychologist is known as offender profiling, or criminal profiling. The practice started during the 1940s during World War II. Today, organizations such as the FBI utilize offender profiling to help apprehend violent criminals. The goal of criminal profiling is to provide law enforcement with a psychological assessment of the suspect and to provide strategies and suggestions that can be used in the interviewing process.

So is the job really as dramatic and exiting as it is portrayed on tv dramas like *Criminal Minds*? "*Criminal Minds* portrays the psychologist as having a more active role than they really do," explained Marc T. Zucker, academic chair of the undergraduate School of Criminal Justice at Kaplan University, in one article. "We all love the thrill of the chase and arrest, however, psychologists don't typically accompany officers in the apprehension of suspects. In addition, many cases take weeks, months or even years to solve, and very rarely are these cases as easy to piece together as they are on the show."

While the job may not be exactly like you see it portrayed on television, the realities of the job are far from boring. Dr. Keith Durkin, chair of the department of psychology and sociology at Ohio Northern University explains, "Careers in criminal psychology are never boring, and if you have an education in that field, it's great training for a huge range of jobs. You can do something different every day. You could work in counselling people who have committed crimes and need psychological assessment. Many psychologists are exploring computer-related fields, like studying Internet predators or helping investigate online fraud." Where Does a Criminal Psychologist Work?: Many people who work in this field spend a great deal of time in office and court settings. A criminal psychologist might spend a considerable amount of time interviewing people, researching an offender's life history or providing expert testimony in the courtroom.

In some cases, criminal psychologists may work closely with police and federal agents to help solve crimes, often by developing profiles of murderers, kidnappers, rapists and other violent individuals. Criminal psychologists are employed in a number of settings. Some work for local, state or federal government, while others are self-employed as independent consultants. In addition to working directly with law enforcement and the courts, criminal psychology may also be employed as private consultants. Still others opt to teach criminal psychology at the university level or at a specialized criminology training facility.

What Training is Needed to Become a Criminal Psychologist?: In many cases, criminal psychologists start out by earning a bachelor's degree in psychology. After completing an undergraduate degree, some students opt to then enter a master's in psychology program. While there are some jobs in criminal and forensic psychology at the master's level, the U.S. Department of Labour reports that opportunities are limited and competition for these positions is often very fierce. Entering a doctorate program after earning your bachelor's is another option. According to the Occupational *Outlook Handbook*, job openings in this specialty area are more plentiful for those with a Ph.D. or Psy.D. degree in psychology.

In order to become a criminal psychologist, you should seriously consider earning a Ph.D. or Psy.D. degree in clinical or counselling psychology. In some cases, students opt to focus on a particular specialty area such as forensic or criminal psychology. The Ph.D (or Doctor of Philosophy) degree is typically more focused on theory and research, while the Psy.D. (or Doctor of Psychology) tends to be more practice-oriented. No matter what type of degree you choose to earn, it will likely take about five years to complete and will include classroom work, practical training, research and a dissertation. In order to become a licensed psychologist, you will also need to complete an internship and pass state examinations. How Much Do Criminal Psychologists Earn?: The typical salaries for criminal psychologists can vary depending on where they work and how much experience they have. According to Payscale.com, national salaries for criminal or forensic psychologists range from a low of $33,900 to a high of $103,000.

Payscale.com also reports that in 2010 criminal and forensic psychologists working for state and local governments, private practice, companies and hospitals tended to have slightly higher average salaries, while those employed by the federal government and non-profit organizations tended to have slightly lower annual salaries. Is Criminal Psychology Right For You? Before you decide if this is the right specialty area for you, spend some time considering your own capabilities and goals. Due to the nature of this profession, you may find yourself dealing with some truly disturbing situations. As a criminal psychologist, you may be called on to look at crime scene photos or interview suspects who may have committed horrifying crimes. Because of this, you need to be prepared to deal with the emotional distress that this type of work may cause. One of the best ways to determine if this career is right for you is to talk to an actual criminal psychologist about what the job is like. Contact your local law enforcement department to see if they can connect you with a criminal psychologist in your area.

## CRIME AND CRIMINOLOGY

The crime according to the social theories is correlated with structure of the society of society. It is also originate from the social disorganization of the society, more urbanization more the ecological problems result in mental strains causes gangs and delinquent crimes. Social conflict theory portrayed crime as a function of social demoralization and a collapse of people's humanity reflecting a decline in society. The brutality of the capitalist system turns workers into animal-like creatures without a will of their own. The Marxist theory of the criminology believe the empirical relation of the development of the capital economy leads to development of "CLASS STRUGGLE" results in indulgence in crime for securing their right from privileged section of the society. The economic inequality intensifies personal problems and crime.

Crime is a product of society and each society will produce its own types and amounts of crime. The critique of Marxist says that if the Marxist is accepted as then what is cause of crime in socialist countries? Marxist standards are too high and moralistic. Powerlessness increases the likelihood of Victimization for women, Crime is a way of "doing gender" for men. The patriarchal structure of the society leads to dominance of the male gender weakening of the position of the woman and causes crimes like cool blood murders, murder in the name of "AZHAT", rapes and prostitution etc. and the patriarchal structure of the society is being resisted by various quarters' causes feminist crime.

The inferior biological structure coupled by weakness of basic instinct like possession, sex and fear of body of human accentuate delinquent behaviour and the tendency towards crimes, and the biological school of thought believe in proper physical development for curbing committing of the crime. Basic theme is that criminals represent physically different structure from non-criminals and criminals as atavistic e.g. sloping foreheads, joined eyebrows, long arms, twisted noses etc. Chambliss and Seidman propounded the modern conflict and on empirical relation of Law, Order and Power in 1971. They believe that the justice system operates to protect the rich and powerful by defining crimes and law enforcement and punishment for law-breaker.

## How to Prevent them?

There are various theories has propounded by jurists for prevention of the crime in societies.

## Punishment

The Marxist has labelled the tool of punishment by saying "only crimes of the poor are punished" and also believes in fact that the crime will virtually disappear with equal distribution of property. The punishment is inflicted for the purposes of breaking norms of the society which has been codified into the law provisions.

## Methods of Punishment

There are various method of punishment has prescribed, radical and classical school of thought extend their appreciation for the strict punishment for curbing the committing of crimes while modern jurist prefer the reformative schools.

## Justifying the Punishment

Cesare Beccaria, follower of classical school, while justifying punishment say that, the retributive approach maintains that punishment should be equal to the harm done, either literally an eye for an eye, or more figuratively which allows for alternative forms of compensation. The retributive approach tends to be retaliatory and vengeance-oriented. The second approach is utilitarian which maintains that punishment should increase the total amount of happiness in the world. This often involves punishment as a means of reforming the criminal, incapacitating him from repeating his crime, and deterring others. Beccaria clearly takes a utilitarian stance. For Beccaria, the purpose of punishment is to create a better society, not revenge. Punishment serves to deter others from committing crimes, and to prevent the criminal from repeating his crime.

## The Nature of Criminology

Criminology is an interdisciplinary study of the making of the law, breaking of the law, and enforcement of the law. Its subject matter covers all topics related to crime and criminal justice. This theory and research on the explanation and causes of crime and criminal delinquency, the rates, distribution, and changes in crime and delinquency in society; individual criminal and delinquent behaviour; criminal victimization and fear of crime.

## History of Criminology

Origin of Criminology is dating back roughly a hundred years when criminologists received their academic training in the social sciences, usually taking degrees in sociology. Criminology gained its place in America academic earlier 20s when sociology was recognized as the home of academic criminology and it gained its knowledge and benefit from the knowledge and insights of the those with little or no academic training but plenty of experience with some aspects of the crime scene.

Now, Criminology is a detailed academic and established subject of the study, this does not means that criminologist agree definition of the field. Those authors of the criminology texts who offer a definition rarely offer the same one. Don C. Gibbons and Peter Garadbedian discuss the competing value per specific that has shaped criminology over the years. They identify three major perspectives: conservative, liberal-cynical and radical sometime called critical.

## COMPUTATIONAL CRIMINOLOGY

Criminology has a rich history of interest in the spatial distribution of crime and criminal events. The sub-field of environmental criminology stems, in part, from this early ecological tradition. Key concepts include the routine nature of many of our daily activities, and the structured way in which we become aware of, and interact with, our environment. The task of simulating the movement and decision-making processes for our hypothetical offender (agent) is complex, but the main elements can be shown here. Principles of environmental criminology suggest that offenders commit crime in areas with which they are most familiar. If we assume that our agent moves about in space much like most people, the agent is then tied to at least three main classes of activity nodes: home, work and recreation.

The agent often travels between these nodes using familiar pathways; the more often an area is visited, the more knowledge the agent will gain regarding the immediate surrounds for both the nodes and the pathways connecting them. At a general level, each agent has an awareness space that characterizes his or her working knowledge of the environment. Within this general area more specialized knowledge is formed by direct experiences, known as activity spaces.

It is here that criminal opportunities are observed and potentially acted upon. If the agent starts his or her day at home, and then travels to either work (or school), for example, he or she will typically take the most direct and most easily navigated route. Along the way, the agent will take notice of a range of phenomena in his or her activity space. For example, the agent might see a favourite coffee house, or take notice of a particular shopping opportunity along the way to his or her work node. Even if the agent does not stop and interact specifically with potential activity sites, he or she will often remember such sites the next time he or she wants to purchase a cup or coffee or to patronize a particular business.

The same learning process applies to the travel paths to and from other significant activity spaces. Environmental criminologists view the learning and decision-making process for crimes to be much the same as those for non-criminal activities. Criminals are rarely criminal all of the time. Everyone has a degree of criminal potential; what differs is the level of criminal propensity that each person 'brings along with us' on our daily routines. More criminally predisposed agents will respond to observed crime opportunities more frequently than people with low (or very low) crime potentials. What requires modelling is the process by which agents move from node to node, and when, if at all, does he or she act upon an observed criminal opportunity. The process of choosing a target of opportunity involves target templating. Each agent compares potential criminal opportunities to his or her crime template in order to assess the value potential rewards against the risks or degree of energy required for successful execution 11, 3. A typical crime in large cities is that of burglary, otherwise known as break and enter, or "B&E". In the large majority of cases, for the agent to be aware of a potential target, the site must be located within the agent's activity space-which, in turn, is defined by the set of common activity nodes, as discussed earlier.

The burglar would travel from Node 1 (home) to Node 2 (work), as his or her routine may require. Along the way he or she recognizes a residential building that is suggestive of a "good" target, as it "fits" within that particular agent's crime template. Research on burglars 34, 35, 1 suggests that variables such as: property value, lack of occupants or potential witnesses capable of intervening, and obvious entry opportunities, all form 'cues' from which the agent assesses quickly to determine if either thecriminal event will be attempted, or at least investigated further.

## Conservative Criminology

Conservative criminology gained ascendancy in America with early writing of parson, Gibbon and Parmelee, who were among the conservative criminologist such later contribution as harry Barnes and Negley teeters, whose text, New horizons in criminology, because a best-seller....Criminal law a given and is interpreted as the codification of the prevailing moral concepts and faith in ultimate the perfectibility of police ad criminal justice machinery."

## Liberal Cynical Criminology

According to Gibbons and Garabedian, liberal Cynical criminology emerged along with the more the 1940s and 1950s because liberal cynical criminology has dominated the field over the past thirsty years, we might also call it mainstream criminology.

## Liberal Criminology

Liberal criminology retained the emphasis on offenders and their behaviour and attempted to explain crimes in terms of either social structure or social process. Those are three major versions of liberal criminology. Another dimension of the liberal criminology is the liberal criminology which has taken deviance from the culture definition of the criminology. Liberal criminology has become a 'gatekeeper' for state domains of control, the value assumptions of hierarchical authority, of centralized controls and a safety valve and temperature gauge in the limits on how far the state can go. The liberal writings of the various sociologists, criminologists and psychologists are given much attention in criminology which is indicative of the continued fascination with power, control and the models of the mechanical world. Their thinking is that man is the centre of the universe, but that they are the centre of man. They prescribe what is good and acceptable and how the world and life processes should be managed.

## Radical Criminology

The radical criminology rejects the liberal reformism that claims have helped to create probation and parole the juvenile court system. They emphasized on the reformatory schools and half way school. More Modern Cynical liberal criminology shades into the newer brands of the liberal criminology but still crime and criminology as manifestation of the exploitation character of monopoly capitalism. Unless the present political-economic structure of America capitalist society in legal machinery will continue to undermine the interest of the people while consolidating those rule.

One of the first things to note and or understand is that "Marxism", as a theoretical perspective, involves a number of different variants or "ways of seeing" the social world. In this respect, although we often tend to talk about Marxism as if there were only one kind just as we tend to talk about other sociological perspectives-such as Functionalism-as if it only involved one basic set of ideas), it is evident that Marxism, as a school of thought, has been interpreted in a number of different ways by various writers. Approach and methods the criminology for Study of Crimes are following.

## The Biological Approach

The biological theories of criminal behaviour says that it study of criminal in biological perspectives.

## Psychogenic Approach

Psychologist investigators are pursing the psychogenic approach to the criminology behaviour, in which the emphasis is based on linking criminal behaviour to mental state, especially mental evidence disease; mental disorders, pathologies, and emotional problems and they repeatedly assert that crime is outcome of criminal mind. The root cause of the criminal behaviour neither environmental nor biological than question seems to be unclear.

## Multifactor Approach

A long-standing criticism of the earlier bio-organic and psychological approach to crime has much of the work entered around the search for single factor or single set of like factors that could be shown to account for all criminal behaviour. The multifactor approach in criminology grew out of the discrepancies and arguments attending the single-factor tradition of the earlier days and its adherent argument for the approach to crime that would reconcile the disparate orientation and contribute made by a variety of the factors. That underlying assumption was that different crimes are result of different combination of the factors. Whose interests are represented by criminal law?

This is not normative system of study but represent the interest of various segment of the society like interest of majority or interest. Criminology represents the interest of groups in society and interests of either all members of the society; the representation of the interest of criminology entire depends upon the type of the system prevailing in the society.

## BIOLOGICAL THEORIES IN CRIMINOLOGY

The biological theories primarily study the physical constitution and endocrinology. They are a bright example of the theories that has not really got any practical support. The misunderstanding of these theories has caused a stereotype that if a person is a criminal then he was born as a criminal and any steps taken in order to change or to influence them are useless. Biological theories are only a part and one the interpretation of criminality but not the only. In the present time there is no assertive evidence of the fact, that the physical constitution and other biological factors cause criminality. Nevertheless, these theories have a right to exist and there was a lot of important information that was used in terms of the development of criminology as a science.

The most vivid example of the biological determinism is the theory of Cesare Lombroso. Lombroso based his theory on the assumption that criminals have certain physiognomic features or abnormalities. Lombroso wanted to be able to detect future criminals in order to isolate them from the society. This gave criminology a strong push to create new methods of dealing with criminals and prevent crimes. Lombroso implied that prisoners had common facial characteristics. If to exaggerate criminals in Lombroso's theory can be identified through the shape of their skulls, asymmetry of the face and head, large cheekbones, ears and lips, long arms and a twisted nose. Lombroso's theory is the oldest one and it can without a doubt be called the main background data for the whole process of the development of criminology. Lombroso stated that men are more inclined to commit crimes due to the conservatism and the narrow-mindedness of their interests. According to Lombroso women have less social contacts and this is what predetermines their conservatism. This theory of female deviance seems to be very discriminating and not a present-day issue. The evaluation of a person as a "born criminal" basing on his facial features is at least not ethical and rather primitive. Lombroso considered this "born criminals" to be the "atavisms" of the society. All the biological theories are based on the notion that biological markers foreordain criminal behaviour. The core of all these theories is that genetic factors or any abnormalities which are inherited or acquired throughout the life, predispose individuals to the criminal behaviour. Lombroso's theory gave life to probably almost every single biological theory that appeared afterward.

Among the biological theories the XYY syndrome occupies a special place, as it analyzes why males are more often subjects to become criminals due to the presence if an extra Y chromosome. As this syndrome is in the first place associated with the low IQ-level of such males which seems to be a much better explanation. IQ shortage causes the inability to adjust in the contemporary social world that requires constant activity and flexibility. People facing these difficulties are left with no other choice that to turn to criminality. It all starts with low performance at school which results in the awareness of the "irretrievable dissimilitude" from others which later on results in deviance. Also such diseases as Organic Brain Syndrome, ADD or hormonal changes according to the biological theories play a very important role in the inclination to any criminal activities.

Another famous biological theory is the theory of William Sheldon. He based his research and inferences on Kretschmer's constitutional personality. For instance, Sheldon pointed out three main body types which are to explain the potential bent to criminal activity: endomorph, mesomorph and ectomorph.

Endomorph is a type hat is fat and therefore is primarily concerned with consumption. Mesomorph is defined through skinny intelligent introverts and ectomorph characterizes large dynamic people. Sheldon claimed that disproportionately mesomorphic people are more often subjects to criminal behaviour than any other body type. In spite of the variety of biological theories nowadays it is clear that there is no possibility to consider that any person can be a "born criminal", because it is very hard to underestimate the social factors and all the psychological issues connected with them. Lombroso's theory was a good start for criminology but this point of view needed to transform into something new and not one-sided.

## CAUSES OF CRIME

How do some people decide to commit a crime? Do they think about the benefits and the risks? Why do some people commit crimes regardless of the consequences? Why do others never commit a crime, no matter how desperate their circumstances? Criminology is the study of crime and criminals by specialists called criminologists. Criminologists study what causes crime and how it might be prevented. Throughout history people have tried to explain what causes abnormal social behaviour, including crime. Efforts to control "bad" behaviour go back to ancient Babylon's Code of Hammurabi some 3,700 years ago. Later in the seventeenth century European colonists in North America considered crime and sin the same thing. They believed evil spirits possessed those who did not conform to social norms or follow rules. To maintain social order in the settlements, persons who exhibited antisocial behaviour had to be dealt with swiftly and often harshly. By the twenty-first century criminologists looked to a wide range of factors to explain why a person would commit crimes. These included biological, psychological, social, and economic Throughout history people have tried to explain why a person would commit crimes. Some consider a life of crime better than a regular job — at least until they are caught.

Reasons for committing a crime include greed, anger, jealously, revenge, or pride. Some people decide to commit a crime and carefully plan everything in advance to increase gain and decrease risk. These people are making choices about their behaviour; some even consider a life of crime better than a regular job — believing crime brings in greater rewards, admiration, and excitement — at least until they are caught. Others get an adrenaline rush when successfully carrying out a dangerous crime. Others commit crimes on impulse, out of rage or fear. The desire for material gain (money or expensive belongings) leads to property crimes such as robberies, burglaries, white-collar crimes, and auto thefts. The desire for control, revenge, or power leads to violent crimes such as murders, assaults, and rapes. These violent crimes usually occur on impulse or the spur of the moment when emotions run high. Property crimes are usually planned in advance.

## Discouraging the Choice of Crime

The purpose of punishment is to discourage a person from committing a crime. Punishment is supposed to make criminal behaviour less attractive and more risky. Imprisonment and loss of income is a major hardship to many people. Another way of influencing choice is to make crime more difficult or to reduce the opportunities. This can be as simple as better lighting, locking bars on auto steering wheels, the presence of guard dogs, or high technology improvements such as security systems and photographs on credit cards.

A person weighing the risks of crime considers factors like how many police officers are in sight where the crime will take place. Studies of New York City records between 1970 and 1999 showed that as the police force in the city grew, less crime was committed. A change in a city's police force, however, is usually tied to its economic health. Normally as unemployment rises, city revenues decrease because fewer people are paying taxes. This causes cutbacks in city services including the police force. So a rise in criminal activity may not be due to fewer police, but rather rising unemployment. Home security consultant conferring with client. Security systems and guard dogs can make crime more difficult or reduce the opportunities for it to occur.

Another means of discouraging people from choosing criminal activity is the length of imprisonment. After the 1960s many believed more prisons and longer sentences would deter crime. Despite the dramatic increase in number of prisons and imposing mandatory lengthy sentences, however, the number of crimes continued to rise. The number of violent crimes doubled from 1970 to 1998. Property crimes rose from 7.4 million to 11 million, while the number of people placed in state and federal prisons grew from 290,000 in 1977 to over 1.2 million in 1998. Apparently longer prison sentences had little effect on discouraging criminal behaviour.

## Parental Relations

Cleckley's ideas on sociopathy were adopted in the 1980s to describe a "cycle of violence" or pattern found in family histories. A "cycle of violence" is where people who grow up with abuse or antisocial behaviour in the home will be much more likely to mistreat their own children, who in turn will often follow the same pattern.

Children who are neglected or abused are more likely to commit crimes later in life than others. Similarly, sexual abuse in childhood often leads these victims to become sexual predators as adults. Many inmates on death row have histories of some kind of severe abuse. The neglect and abuse of children often progresses through several generations. The cycle of abuse, crime, and sociopathy keeps repeating itself.

Children who are neglected or abused commit substantially more crimes later in life than others. The cycle of violence concept, based on the quality of early life relationships, has its positive counterpart. Supportive and loving parents who respond to the basic needs of their child instill self-confidence and an interest in social environments. These children are generally well-adjusted in relating to others and are far less likely to commit crimes.

By the late twentieth century the general public had not accepted that criminal behaviour is a psychological disorder but rather a willful action. The public cry for more prisons and tougher sentences outweighed rehabilitation and the treatment of criminals. Researchers in the twenty-first century, however, continued to look at psychological stress as a driving force behind some crimes.

## Hormones

Hormones are bodily substances that affect how organs in the body function. Researchers also looked at the relationship between hormones, such as testosterone and cortisol, and criminal behaviour. Testosterone is a sex hormone produced by male sexual organs that cause development of masculine body traits. Cortisol is a hormone produced by adrenal glands located next to the kidneys that effects how quickly food is processed by the digestive system. Higher cortisol levels leads to more glucose to the brain for greater energy, such as in times of stress or danger. Animal studies showed a strong link between high levels of testosterone and aggressive behaviour. Testosterone measurements in prison populations also showed relatively high levels in the inmates as compared to the U.S. adult male population in general. Studies of sex offenders in Germany showed that those who were treated to remove testosterone as part of their sentencing became repeat offenders only 3 percent of the time. This rate was in stark contrast to the usual 46 percent repeat rate. These and similar studies indicate testosterone can have a strong bearing on criminal behaviour.

Cortisol is another hormone linked to criminal behaviour. Research suggested that when the cortisol level is high a person's attention is sharp and he or she is physically active. In contrast, researchers found low levels of cortisol were associated with short attention spans, lower activity levels, and often linked to antisocial behaviour including crime. Studies of violent adults have shown lower levels of cortisol; some believe this low level serves to numb an offender to the usual fear associated with committing a crime and possibly getting caught. It is difficult to isolate brain activity from social and psychological factors, as well as the effects of substance abuse, parental relations, and education. Yet since some criminals are driven by factors largely out of their control, punishment will not be an effective deterrent. Help and treatment become the primary responses.

## Heredity and Brain Activity

Searching for the origins of antisocial personality disorders and their influence over crime led to studies of twins and adopted children in the 1980s. Identical twins have the exact same genetic makeup. Researchers found that identical twins were twice as likely to have similar criminal behaviour than fraternal twins who have similar but not identical genes, just like any two siblings. Other research indicated that adopted children had greater similarities of crime rates to their biological parents than to their adoptive parents. These studies suggested a genetic basis for some criminal behaviour.

Prisoner in California being prepared for a lobotomy in 1961. At the time, many psychiatrists believed that criminal behaviour was lodged in certain parts of the brain, and lobotomies were frequently done on prisoners.

With new advances in medical technology, the search for biological causes of criminal behaviour became more sophisticated. In 1986 psychologist Robert Hare identified a connection between certain brain activity and antisocial behaviour. He found that criminals experienced less brain reaction to dangerous situations than most people. Such a brain function, he believed, could lead to greater risk-taking in life, with some criminals not fearing punishment as much as others.

Studies related to brain activity and crime continued into the early twenty-first century. Testing with advanced instruments probed the inner workings of the brain. With techniques called computerized tomography (CT scans), magnetic resonance imaging (MRI), and positron emission tomography (PET), researchers searched for links between brain activity and a tendency to commit crime. Each of these tests can reveal brain activity.

Research on brain activity investigated the role of neurochemicals, substances the brain releases to trigger body activity, and hormones in influencing criminal behaviour. Studies indicated that increased levels of some neurochemicals, such as serotonin, decreases aggression. Serotonin is a substance produced by the central nervous system that has broad sweeping effects on the emotional state of the individual.

In contrast higher levels of others, such as dopamine, increased aggression. Dopamine is produced by the brain and affects heart rate and blood pressure. Researchers expected to find that persons who committed violent crimes have reduced levels of serotonin and higher levels of dopamine. This condition would have led to periods of greater activity including aggression if the person is prone towards aggression. In the early twenty-first century researchers continued investigating the relationship between neurochemicals and antisocial behaviour, yet connections proved complicated. Studies showed, for example, that even body size could influence the effects of neurochemicals and behaviour.

## Education

Conforming to Merton's earlier sociological theories, a survey of inmates in state prisons in the late 1990s showed very low education levels. Many could not read or write above elementary school levels, if at all. The most common crimes committed by these inmates were robbery, burglary, automobile theft, drug trafficking, and shoplifting.

Because of their poor educational backgrounds, their employment histories consisted of mostly low wage jobs with frequent periods of unemployment. Employment at minimum wage or below living wage does not help deter criminal activity.

Even with government social services, such as public housing, food stamps, and medical care, the income of a minimum wage household still falls short of providing basic needs. People must make a choice between continued long-term low income and the prospect of profitable crime. Gaining further education, of course, is another option, but classes can be expensive and time consuming. While education can provide the chance to get a better job, it does not always overcome the effects of abuse, poverty, or other limiting factors.

## Peer Influence

A person's peer group strongly influences a decision to commit crime. For example, young boys and girls who do not fit into expected standards of academic achievement or participate in sports or social programs can sometimes become Crack cocaine pipe displayed by police. Drugs and alcohol impair judgment and reduce inhibitions, giving a person greater courage to commit a crime. Lost in the competition.

Children of families who cannot afford adequate clothing or school supplies can also fall into the same trap. Researchers believe these youth may abandon schoolmates in favor of criminal gangs, since membership in a gang earns respect and status in a different manner.

In gangs, antisocial behaviour and criminal activity earns respect and street credibility. Like society in general, criminal gangs are usually focused on material gain. Gangs, however, resort to extortion, fraud, and theft as a means of achieving it. The fear of young people, mostly boys, joining gangs influenced many government projects in the last half of the twentieth century including President Lyndon Johnson's (1908–1973; served 1963–69) "War on Crime" programs.

## Drugs and Alcohol

Some social factors pose an especially strong influence over a person's ability to make choices. Drug and alcohol abuse is one such factor. The urge to commit crime to support a drug habit definitely influences the decision process. Both drugs and alcohol impair judgment and reduce inhibitions (socially defined rules of behaviour), giving a person greater courage to commit a crime. Deterrents such as long prison sentences have little meaning when a person is high or drunk.

Substance abuse, commonly involving alcohol, triggers "stranger violence," a crime in which the victim has no relationship whatsoever with his or her attacker. Such an occurrence could involve a confrontation in a bar or some other public place where the attacker and victim happen to be at the same time.

Criminologists estimate that alcohol or drug use by the attacker is behind 30 to 50 percent of violent crime, such as murder, sexual assault, and robbery. In addition drugs or alcohol may make the victim a more vulnerable target for a criminal by being less attentive to activities around and perhaps visiting a poorly lighted or secluded area not normally frequented perhaps to purchase drugs. The idea that drug and alcohol abuse can be a major factor in a person's life is why there are numerous treatment programs for young people addicted to these substances. Treatment focuses on positive support to influence a person's future decision making and to reduce the tendency for antisocial and criminal behaviour.

## Easy Access

Another factor many criminologists consider key to making a life of crime easier is the availability of handguns in U.S. society. Many firearms used in crimes are stolen or purchased illegally (bought on what is called the "black market"). Firearms provide a simple means of committing a crime while allowing offenders some distance or detachment from their victims. Of the 400,000 violent crimes involving firearms in 1998, over 330,000 involved handguns. By the beginning of the twenty-first century firearm use was the eighth leading cause of death in the United States.

Similarly, the increased availability of free information on the Internet also makes it easy to commit certain kinds of At the beginning of the twenty-first century, firearm use was the eighth leading cause of death in the United States. (AP/Wide World Photos) crime. Web sites provide instructions on how to make bombs and buy poisons; all this information is easily available from the comfort of a person's home. Easy access, however, will not be the primary factor in a person's decision to commit a crime. Other factors—biological, psychological, or social—will also come into play.

## HEREDITY AND BRAIN ACTIVITY

Searching for the origins of antisocial personality disorders and their influence over crime led to studies of twins and adopted children in the 1980s. Identical twins have the exact same genetic makeup.

Researchers found that identical twins were twice as likely to have similar criminal behaviour than fraternal twins who have similar but not identical genes, just like any two siblings. Other research indicated that adopted children had greater similarities of crime rates to their biological parents than to their adoptive parents. These studies suggested a genetic basis for some criminal behaviour.

With new advances in medical technology, the search for biological causes of criminal behaviour became more sophisticated. In 1986 psychologist Robert Hare identified a connection between certain brain activity and antisocial behaviour. He found that criminals experienced less brain reaction to dangerous situations than most people. Such a brain function, he believed, could lead to greater risk-taking in life, with some criminals not fearing punishment as much as others.

Studies related to brain activity and crime continued into the early twenty-first century. Testing with advanced instruments probed the inner workings of the brain. With techniques called computerized tomography (CT scans), magnetic resonance imaging (MRI), and positron emission tomography (PET), researchers searched for links between brain activity and a tendency to commit crime. Each of these tests can reveal brain activity.

Research on brain activity investigated the role of neurochemicals, substances the brain releases to trigger body activity, and hormones in influencing criminal behaviour. Studies indicated that increased levels of some neurochemicals, such as serotonin, decreases aggression. Serotonin is a substance produced by the central nervous system that has broad sweeping effects on the emotional state of the individual.

In contrast higher levels of others, such as dopamine, increased aggression. Dopamine is produced by the brain and affects heart rate and blood pressure. Researchers expected to find that persons who committed violent crimes have reduced levels of serotonin and higher levels of dopamine. This condition would have led to periods of greater activity including aggression if the person is prone towards aggression.

In the early twenty-first century researchers continued investigating the relationship between neurochemicals and antisocial behaviour, yet connections proved complicated. Studies showed, for example, that even body size could influence the effects of neurochemicals and behaviour.

## Hormones

Hormones are bodily substances that affect how organs in the body function. Researchers also looked at the relationship between hormones, such as testosterone and cortisol, and criminal behaviour. Testosterone is a sex hormone produced by male sexual organs that cause development of masculine body traits. Cortisol is a hormone produced by adrenal glands located next to the kidneys that effects how quickly food is processed by the digestive system. Higher cortisol levels leads to more glucose to the brain for greater energy, such as in times of stress or danger. Animal studies showed a strong link between high levels of testosterone and aggressive behaviour. Testosterone measurements in prison populations also showed relatively high levels in the inmates as compared to the U.S. adult male population in general. Studies of sex offenders in Germany showed that those who were treated to remove testosterone as part of their sentencing became repeat offenders only 3 percent of the time. This rate was in stark contrast to the usual 46 percent repeat rate. These and similar studies indicate testosterone can have a strong bearing on criminal behaviour. Cortisol is another hormone linked to criminal behaviour. Research suggested that when the cortisol level is high a person's attention is sharp and he or she is physically active. In contrast, researchers found low levels of cortisol were associated with short attention spans, lower activity levels, and often linked to antisocial behaviour including crime.

Studies of violent adults have shown lower levels of cortisol; some believe this low level serves to numb an offender to the usual fear associated with committing a crime and possibly getting caught. It is difficult to isolate brain activity from social and psychological factors, as well as the effects of substance abuse, parental relations, and education. Yet since some criminals are driven by factors largely out of their control, punishment will not be an effective deterrent. Help and treatment become the primary responses.

## Education

Conforming to Merton's earlier sociological theories, a survey of inmates in state prisons in the late 1990s showed very low education levels. Many could not read or write above elementary school levels, if at all.

The most common crimes committed by these inmates were robbery, burglary, automobile theft, drug trafficking, and shoplifting. Because of their poor educational backgrounds, their employment histories consisted of mostly low wage jobs with frequent periods of unemployment.

Employment at minimum wage or below living wage does not help deter criminal activity. Even with government social services, such as public housing, food stamps, and medical care, the income of a minimum wage household still falls short of providing basic needs. People must make a choice between continued long-term low income and the prospect of profitable crime. Gaining further education, of course, is another option, but classes can be expensive and time consuming. While education can provide the chance to get a better job, it does not always overcome the effects of abuse, poverty, or other limiting factors.

## Easy Access

Another factor many criminologists consider key to making a life of crime easier is the availability of handguns in U.S. society. Many firearms used in crimes are stolen or purchased illegally (bought on what is called the "black market").

Firearms provide a simple means of committing a crime while allowing offenders some distance or detachment from their victims. Of the 400,000 violent crimes involving firearms in 1998, over 330,000 involved handguns. By the beginning of the twenty-first century firearm use was the eighth leading cause of death in the United States.

## CRIMES AGAINST WOMEN

Police records show high incidence of crimes against women in India. The National Crime Records Bureau reported in 1998 that the growth rate of crimes against women would be higher than the population growth rate by 2010. Earlier, many cases were not registered with the police due to the social stigma attached to rape and molestation cases. Official statistics show that there has been a dramatic increase in the number of reported crimes against women.

## Dowry

In 1961, the Government of India passed the Dowry Prohibition Act, making the dowry demands in wedding arrangements illegal. However, many cases of dowry-related domestic violence, suicides and murders have been reported. In the 1980s, numerous such cases were reported. However, recent reports show that the number of these crimes have reduced drastically. In 1985, the Dowry Prohibition (maintenance of lists of presents to the bride and bridegroom) rules were framed. According to these rules, a signed list of presents given at the time of the marriage to the bride and the bridegroom should be maintained. The list should contain a brief description of each present, its approximate value, the name of whoever has given the present and his/her relationship to the person. A 1997 report claimed that at least 5,000 women die each year because of dowry deaths, and at least a dozen die each day in 'kitchen fires' thought to be intentional. The term for this is "bride burning" and is criticized within India itself. Amongst the urban educated, such dowry abuse has reduced dramatically.

## Child Marriage

Child marriage has been traditionally prevalent in India and continues to this day. Young girls live with their parents until they reach puberty. In the past, the child widows were condemned to a life of great agony, shaving heads, living in isolation, and shunned by the society. Although child marriage was outlawed in 1860, it is still a common practice. According to UNICEF's "State of the World's Children-2009" report, 47% of India's women aged 20-24 were married before the legal age of 18, with 56% in rural areas. The report also showed that 40% of the world's child marriages occur in India.

## Female Infanticides and Sex Selective Abortions

India has a highly masculine sex ratio, the chief reason being that many women die before reaching adulthood. Tribal societies in India have a less masculine sex ratio than all other caste groups. This, in spite of the fact that tribal communities have far lower levels of income, literacy and health facilities. It is therefore suggested by many experts, that the highly masculine sex ratio in India can be attributed to female infanticides and sex-selective abortions.

All medical tests that can be used to determine the sex of the child have been banned in India, due to incidents of these tests being used to get rid of unwanted female children before birth. Female infanticide (killing of girl infants) is still prevalent in some rural areas. The abuse of the dowry tradition has been one of the main reasons for sex-selective abortions and female infanticides in India.

## Domestic Violence

The incidents of domestic violence are higher among the lower Socio-Economic Classes (SECs). There are various instances of an inebriated husband beating up the wife often leading to severe injuries. Domestic violence is also seen in the form of physical abuse. The Protection of Women from Domestic Violence Act, 2005 came into force on October 26, 2006.

## Sexual Harassment

Half of the total number of crimes against women reported in 1990 related to molestation and harassment at the workplace. Eve teasing is a euphemism used for sexual harassment or molestation of women by men. Many activists blame the rising incidents of sexual harassment against women on the influence of "Western culture". In 1987, The Indecent Representation of Women (Prohibition) Act was passed to prohibit indecent representation of women through advertisements or in publications, writings, paintings, figures or in any other manner. In 1997, in a landmark judgement, the Supreme Court of India took a strong stand against sexual harassment of women in the workplace. The Court also laid down detailed guidelines for prevention and redressal of grievances. The National Commission for Women subsequently elaborated these guidelines into a Code of Conduct for employers.

While public urination is not practised by men of all ages in India, it is socially unacceptable for girls and women to publicly urinate when restrooms are unavailable. In other countries such as Laos, Cambodia, and Vietnam public urination is practised by women when there are no toilets. This leads to harassment and UTI in women.

## Trafficking

The Immoral Traffic (Prevention) Act was passed in 1956. Cite error: Closing <ref> missing for <ref> tag. Because of such geographical location, India experiences large amount of drug trafficking through the borders. India is the world's largest producer of licit opium.

But opium is diverted to illicit international drug markets. India is a transshipment point for heroin from Southwest Asian countries like Afghanistan and Pakistan and from Southeast Asian countries like Burma, Laos, and Thailand. Heroin is smuggled from Pakistan and Burma, with some quantities transsshipped through Nepal. Most heroin shipped from India are destined for Europe. There have been reports of heroin smuggled from Mumbai to Nigeria for further export. In Maharashtra, Mumbai is an important centre for distribution of drug. The most commonly used drug in Mumbai is Indian heroin (called *desi mal* by the local population). Both public transportation (road and rail transportation) and private transportation are used for this drug trade.

Drug trafficking affects the country in many ways.
- Drug abuse: Cultivation of illicit narcotic substances and drug trafficking affects the health of the individuals and destroy the economic structure of the family and society.
- Organized crime: Drug trafficking results in growth of organized crime which affects social security. Organised crime connects drug trafficking with corruption and money laundering.
- Political instability: Drug trafficking also aggravate the political instability in North-West and North-East India.

A survey conducted in 2003-2004 by Narcotics Control Bureau found that India has at least four million drug addicts. The most common drugs used in India are cannabis, hashish, opium and heroin. In 2006 alone, India's law enforcing agencies recovered 230 kg heroin and 203 kg of cocaine. In an annual government report in 2007, the United States named India among 20 major hubs for trafficking of illegal drugs along with Pakistan, Afghanistan and Burma. However, studies reveal that most of the criminals caught in this crime are either Nigerian or US nationals.

Several measures have been taken by the Government of India to combat drug trafficking in the country. India is a party of the Single Convention on Narcotic Drugs (1961), the Convention on Psychotropic Substances (1971), the Protocol Amending the Single Convention on Narcotic Drugs (1972) and the United Nations Convention Against Illicit Traffic in Narcotic Drugs and Psychotropic Substances (1988).

An Indo-Pakistani committee was set up in 1986 to prevent trafficking in narcotic drugs. India signed a convention with the United Arab Emirates in 1994 to control drug trafficking. In 1995, India signed an agreement with Egypt for investigation of drug cases and exchange of information and a Memorandum of Understanding of the Prevention of Illicit Trafficking in Drugs with Iran.

## Arms Trafficking

According to a joint report published by Oxfam, Amnesty International and the International Action Network on Small Arms (IANSA) in 2006, there are around 40 million illegal small arms in India out of approximately 75 million in worldwide circulation. Majority of the illegal small arms make its way into the states of Bihar, Chhattisgarh, Uttar Pradesh, Jharkhand, Orissa and Madhya Pradesh.

In India, a used AK-47 costs $3,800 in black market. Large amount of illegal small arms are manufactured in various illegal arms factories in Uttar Pradesh and Bihar and sold on the black market for as little as $5.08. Chinese pistols are in demand in the illegal small arms market in India because they are easily available and cheaper. This trend poses a significant problem for the states of Bihar, Uttar Pradesh, Jharkhand, Chhattisgarh, Orissa, Maharashtra, West Bengal, Karnataka and Andhra Pradesh which have influence of Naxalism. The porous Indo-Nepal border is an entry point for Chinese pistols, AK-47 and M-16 rifles into India as these arms are used by the Naxalites who have ties to Maoists in Nepal.

In North-East India, there is a huge influx of small arms due to the insurgent groups operating there. The small arms in North-East India come from insurgent groups in Burma, black market in South-East Asian countries like Pakistan, Bangladesh, Nepal and Sri Lanka, black market in Cambodia, the People's Republic of China, insurgent groups like the Liberation Tigers of Tamil Eelam, the Communist Party of India (Maoist), the Communist Party of Nepal (Maoist), Indian states like Uttar Pradesh and pilferages from legal gun factories, criminal organizations operating in India and South Asian countries and other international markets like Romania, Germany etc. The small arms found in North-East India are M14 rifle, M16 rifle, AK-47, AK-56, AK-74, light machine guns, Chinese hand grenades, mines, rocket-propelled grenades, submachine guns etc. The Ministry of External Affairs and Ministry of Home Affairs drafted a joint proposal to the United Nations, seeking a global ban on small-arms sales to non-state users.

## CONSEQUENCES OF CRIMES

Emotional distress as the result of crime is a recurring theme for all victims of crime. The most common problem, affecting three quarters of victims, were psychological problems, including: fear, anxiety, nervousness, self-blame, anger, shame, and difficulty sleeping. These problems often result in the development of chronic PTSD (post-traumatic stress disorder). Post crime distress is also linked to pre-existing emotional problems and sociodemographic variables. This has known to become a leading case of the elderly to be more adversely affected.

Victims may experience the following psychological reactions:
- Increase in the belief of personal vulnerability.
- The perception of the world as meaningless and incomprehensible.
- The view of themselves in a negative light.

The experience of victimization may result in an increasing fear of the victim of the crime, and the spread of fear in the community.

### Victim Proneness

One of the most controversial sub-topics within the broader topic is victim proneness. The concept of victim proneness is a "highly moralistic way of assigning guilt" to the *victim of a crime*, also known as victim blaming. One theory, the *environmental theory*, posits that the location and context of the crime gets the victim of the crime and the perpetrator of that crime together..

There have been some studies recently to quantify the real existence of victim-proneness. Contrary to the urban legend that more women are repeat victims, and thus more victim-prone than men, actually men in their prime (24 to 34 year old males) are more likely to be victims of repeated crimes. While each study used different methodology, their results must be taken seriously and further studies are warranted.

The study of victimology may also include the "culture of victimhood," wherein the victim of a crime revels in his status, proclaiming that self-created victimhood throughout a community by winning the sympathy of professionals and peers.

In the case of juvenile offenders, the study results also show that people are more likely to be victimized as a result of a serious offence by someone they know; the most frequent crimes committed by adolescents towards someone they know were sexual assault, common assault, and homicide. Adolescents victimizing people they did not know generally committed common assault, forcible confinement, armed robbery, and robbery

## Victim Facilitation

Victim facilitation, another controversial sub-topic, but a more accepted theory than victim blaming, finds its roots in the writings of criminologists Marvin Wolfgang. The choice to use victim facilitation as opposed to "victim blaming" or some other term is that victim facilitation is not blaming the victim, but rather the interactions of the victim that make he/she vulnerable to a crime. While victim facilitation relates to "victim blaming" the idea behind victim facilitation is to study the elements that make a victim more accessible or vulnerable to an attack. In an article that summarizes the major movements in victimology internationally, Schneider expresses victim facilitation as a model that ultimately describes only the misinterpretation of victim behaviour of the offender.

It is based upon the theory of a symbolic interaction and does not alleviate the offender of his/her exclusive responsibility. In Eric Hickey's *Serial Murderers and their Vicitms,* a major analysis of 329 serial killers in America is conducted. As part of Hickey's analysis, he categorized victims as high, low, or mixed regarding the victim's facilitation of the murder.

Categorization was based upon lifestyle risk (example, amount of time spent interacting with strangers), type of employment, and their location at the time of the killing (example, bar, home or place of business).

Hickey found that 13-15% of victims had high facilitation, 60-64% of victims had low facilitation and 23-25% of victims had a combination of high and low facilitation. Hickey also noted that among serial killer victims after 1975, one in five victims placed themselves at risk either by hitchhiking, working as a prostitute or involving themselves in situations in which they often came into contact with strangers.

There is importance in studying and understanding victim facilitation as well as continuing to research it as a sub-topic of victimization. For instance, a study of victim facilitation increases public awareness, leads to more research on victim-offender relationship, and advances theoretical etiologies of violent crime.

One of the ultimate purposes of this type of knowledge is to inform the public and increase awareness so less people become victims. Another goal of studying victim facilitation, as stated by Godwin, is to aid in investigations.

Godwin discusses the theory of victim social networks as a concept in which one looks at the areas of highest risk for victimization from a serial killer. This can be connected to victim facilitation because the victim social networks are the locations in which the victim is most vulnerable to the serial killer. Using this process, investigators can create a profile of places where the serial killer and victim both frequent.

## Studies

The study of victims is multidisciplinary. It does not just cover victims of crime, but also victims of (traffic) accidents, natural disasters, war crimes and abuse of power. The professionals involved in victimology may be scientists, practitioners and policy makers. Studying victims can be done from the perspective of the individual victim, a third person, or from an epidemiological point of view.

## Victimization Rate in United States

The National Crime Victimization Survey (NCVS) is a tool to measure the existence of actual, rather than reported crimes—the victimization rate. The National Crime Victimization Survey is the United States': "primary source of information on crime victimization. Each year, data are obtained from a nationally represented sample of 77,200 households comprising nearly 134,000 persons on the frequency, characteristics and consequences of criminal victimization in the United States. This survey enables the (government) to estimate the likelihood of victimization by rape, sexual assault, robbery, assault, theft, household burglary, and motor vehicle theft for the population as a whole as well as for segments of the population such as women, the elderly, members of various racial groups, city dwellers, or other groups." According to the Bureau of Justice Statistics (BJS), the NCVS reveals that, from 1994 to 2005, violent crime rates have declined, reaching the lowest levels ever recorded. Property crimes continue to decline.

## International Crime Victims Survey

Many countries have such victimization surveys. They give a much better account for the volume crimes but are less accurate for crimes that occur with a (relative) low frequency such as homicide, or victimless 'crimes' such as drug (ab)use. Attempts to use the data from these national surveys for international comparison have failed. Differences in definitions of crime and other methodological differences are too big for proper comparison. A dedicated survey for international comparison: A group of European criminologists started an international victimization study with the sole purpose to generate international comparative crime and victimization data. The project is now known as the International Crime Victims Survey (ICVS). After the first round in 1989, the surveys were repeated in 1992, 1996, and 2000 and 2004/2005.

## Society as Crime Victim

There is one strain of thought that society itself is the victim of many crimes, especially such homicide felonies as murder and manslaughter. This sentiment has been espoused by many lawyers, judges, and academics. Some district attorneys feel they represent all of society, while some feel they are the lawyer for the crime victim. Say there was a serious crime, a murder. You read about it in the news. You are definitely not a victim, but it has affected your life in various ways. Maybe you will be more cautious, less trusting, and show a colder shoulder to strangers. Your attitude affects everyone you come across in your daily life, and it has repercussions on all of them. There you have it. All of them are part of society and this shall spread. As contradictory as it may seem, the victim and his/her closed ones also may blame society. Where was the police? Where was help and help was most needed? Where was the good samaritan? Where was anyone? Please! Please? Please! Finally, where was God?

## Penal Couple

The *penal couple* is defined as the relationship between perpetrator and victim of a crime. A sociologist invented the term in 1963. The term is now accepted by many sociologists. The concept is, essentially, that "when a crime takes place, it has two partners, one the offender and second the victim, who is providing opportunity to the criminal in committing the crime." The victim, in this view, is "a participant in the penal couple and should bear some 'functional responsibility' for the crime." The very idea is strongly rejected by some other victimologists as blaming the victim.

## Rights of Victims

In 1985, the UN General Assembly adopted the Declaration on the Basic Principles of Justice for Victims of Crime and Abuse of Power. Also, the International Victimology Institute Tilburg (INTERVICT) and the World Society of Victimology developed a UN Convention for Victims of Crime and Abuse of Power.

## DISCOURAGING THE CHOICE OF CRIME

The purpose of punishment is to discourage a person from committing a crime. Punishment is supposed to make criminal behaviour less attractive and more risky. Imprisonment and loss of income is a major hardship to many people. Another way of influencing choice is to make crime more difficult or to reduce the opportunities. This can be as simple as better lighting, locking bars on auto steering wheels, the presence of guard dogs, or high technology improvements such as security systems and photographs on credit cards. A person weighing the risks of crime considers factors like how many police officers are in sight where the crime will take place. Studies of New York City records between 1970 and 1999 showed that as the police force in the city grew, less crime was committed. A change in a city's police force, however, is usually tied to its economic health. Normally as unemployment rises, city revenues decrease because fewer people are paying taxes. This causes cutbacks in city services including the police force. So a rise in criminal activity may not be due to fewer police, but rather rising unemployment.

Home security consultant conferring with client. Security systems and guard dogs can make crime more difficult or reduce the opportunities for it to occur. (Ms. Martha Tabor/Working Images Photographs) Another means of discouraging people from choosing criminal activity is the length of imprisonment. After the 1960s many believed more prisons and longer sentences would deter crime. Despite the dramatic increase in number of prisons and imposing mandatory lengthy sentences, however, the number of crimes continued to rise.

The number of violent crimes doubled from 1970 to 1998. Property crimes rose from 7.4 million to 11 million, while the number of people placed in state and federal prisons grew from 290,000 in 1977 to over 1.2 million in 1998. Apparently longer prison sentences had little effect on discouraging criminal behaviour.

## Parental Relations

Cleckley's ideas on sociopathy were adopted in the 1980s to describe a "cycle of violence" or pattern found in family histories. A "cycle of violence" is where people who grow up with abuse or antisocial behaviour in the home will be much more likely to mistreat their own children, who in turn will often follow the same pattern.

Children who are neglected or abused are more likely to commit crimes later in life than others. Similarly, sexual abuse in childhood often leads these victims to become sexual predators as adults. Many inmates on death row have histories of some kind of severe abuse. The neglect and abuse of children often progresses through several generations. The cycle of abuse, crime, and sociopathy keeps repeating itself. Children who are neglected or abused commit substantially more crimes later in life than others. The cycle of violence concept, based on the quality of early life relationships, has its positive counterpart. Supportive and loving parents who respond to the basic needs of their child instill self-confidence and an interest in social environments.

These children are generally well-adjusted in relating to others and are far less likely to commit crimes.

By the late twentieth century the general public had not accepted that criminal behaviour is a psychological disorder but rather a wilful action. The public cry for more prisons and tougher sentences outweighed rehabilitation and the treatment of criminals. Researchers in the twenty-first century, however, continued to look at psychological stress as a driving force behind some crimes.

## CRIMINAL PROFILING FROM CRIME SCENE ANALYSIS

Criminal profiling has been used successfully by law enforcement in several areas and is a valued means by which to narrow the field of investigation. Profiling does not provide the specific identity of the offender. Rather, it indicates the kind of person most likely to have committed a crime by focusing on certain behavioural and personality characteristics.

Profiling techniques have been used in various settings such as hostage taking. Law enforcement officers need to learn as much as possible about the hostage taker in order to protect the lives of the hostages. In such cases, police are aided by verbal contact (although often limited) with the offender and possibly by access to his family and friends. They must be able to assess the subject in terms of what course of action he is likely to take and what his reactions to various stimuli might be.

Profiling has been used also in identifying anonymous letter writers and persons who make written or spoken threats of violence. In cases of the latter psycholinguistic techniques have been used to compose a "threat dictionary." whereby every word in a message is assigned by computer, to a specific category. Words as they are used in the threat message are then compared with those words as they are used in ordinary speech or writings. The vocabulary usage in the message may yield "signature" words unique to the offender. In this way, police may not only be able to determine that several letters were written by the same individual, but also to learn about the background and psychology of the offender.

Rapists and arsonists also lend themselves to profiling techniques. Through careful interview of the rape victim about the rapist's behaviour law enforcement personnel begin to build a profile of the offender. The rationale behind this approach is that behaviour reflects personality, and by examining behaviour the investigator may be able to determine what type of person is responsible for the offense. For example, common characteristics of arsonists have been derived from an analysis of the data from the FBI's Crime in the United States. Knowledge of these characteristics can aid the investigator in identifying possible suspects and in developing techniques and strategies for interviewing them. However, studies in this area have focused on specific categories of offenders and are not yet generalizable to all offenders.

Criminal profiling has been found to be of particular usefulness in crimes such as serial sexual homicides. These crimes create a great deal of fear because of their apparently random and motiveless nature, and they are also given high publicity. Consequently law enforcement personnel are under great public pressure to apprehend the perpetrator as quickly as possible. At the same time, these crimes may be the most difficult to solve, precisely because of their apparent randomness. While it is not completely accurate to say that these crimes are motiveless, the motive may all too often be one understood only by the perpetrator. Lunde (1976) demonstrates this issue in terms of the victims chosen by a particular offender. As Lunde points out, although the serial murderer may not know his victims their selection is not random. Rather, it is based on the murderer's perception of certain characteristics of his victims that are of symbolic significance to him. An analysis of the similarities and differences among victims of a particular serial murderer provides important information concerning the "motive" in an apparently motiveless crime. This in turn, may yield information about the perpetrator himself. For example the murder may be the result of a sadistic fantasy in the mind of the murderer and a particular victim may be targeted because of a symbolic aspect of the fantasy.

In such cases, the investigating officer faces a completely different situation from the one in which a murder occurs as the result of jealousy or a family quarrel, or during the commission of another felony. In those cases, a readily identifiable motive may provide vital clues about the identity of the perpetrator. In the case of the apparently motiveless crime, law enforcement may need to look to other methods in addition to conventional investigative techniques, in its efforts to identify the perpetrator. In this context, criminal profiling has been productive, particularly in those crimes where the offender has demonstrated repeated patterns at the crime scene.

## The Profiling of Murderers

Traditionally two very different disciplines have used the technique of profiling murderers: mental health clinicians who seek to explain the personality and actions of a criminal through psychiatric concepts, and law enforcement agents whose task is to determine the behavioural patterns' of a suspect through investigative concepts.

# 5

# Psychological Testing

Psychological testing is a field characterized by the use of samples of behaviour in order to assess psychological construct(s), such as cognitive and emotional functioning, about a given individual. The technical term for the science behind psychological testing is psychometrics. By *samples of behaviour*, one means observations of an individual performing tasks that have usually been prescribed beforehand, which often means scores on a test. These responses are often compiled into statistical tables that allow the evaluator to compare the behaviour of the individual being tested to the responses of a norm group.

## Current Status

The status of psychological testing can best be described as extremely ambiguous. This is true despite the fact that Buros (1974), the editor of *Tests in Print*, listed 2,476 tests that were available in 1974. However, in many respects psychological testing has been on a downhill course in the last two decades. Bersoff (1973) refers to this decline as turning "a silk purse into a sow's ear." He notes:

For almost 50 years, beginning with World War I, psychological testing was perceived as the vehicle by which major decisions about people's lives could be made in industry, the military, hospitals, mental health clinics, and the schools. Scores derived from psychometric instruments were used to classify, segregate, track, advance, employ, institutionalise, and educate people. Now, IQ testing is outlawed in San Francisco, personnel selection tests are declared illegal unless directly relevant to employment, group intelligence measures are banned in New York City schools, a whole profession which has distinguished itself from psychiatry primarily because its practitioners can test has been declared moribund, and school psychologists in Boston have been declared incompetent. In the last 10 years, what was once a silk purse has been transformed into a sow's ear.

Who is to be held accountable for this psychological alchemy? The answer is two brands of "psychos": psychoanalysts and psychometricians.

Psychoanalysts are to blame because they have perpetrated a fraudulent (Freudulent?) theory of personality and have perpetuated its myth. Psychometrists, the test constructors, are to blame because they have forgotten their historical antecedents and have become overly concerned with psychometric aesthetics to the neglect of validity.

The above quotation captures only *some* of the problems in psychological testing and obviously reflects the author's own biases. Although we are generally in sympathy with Bersoff's position, a more objective way to evaluate testing is to look at surveys conducted over the last two decades surveys on the use of and attitudes toward testing held by clinical psychologists. In addition to determining the popularity of various tests, these surveys have examined the discrepant values ascribed to testing by both academic and clinical psychologists practicing in the community and in institutional settings. The decrease in publishing related to psychological testing has also been documented.

Between 1961 and 1976 three surveys were conducted to determine how psychological tests were used in the United States. The general similarity of these surveys allows us to trace changes in test usage over this fifteen-year period. The questionnaires developed for these surveys listed commonly used psychological tests, allowing the respondent to indicate whether the test was used in the first place, and if so, how frequently (i.e., 0 = never, 1 = occasionally, 2 = frequently, 3 = majority of cases). In each study, an attempt was made to send questionnaires to a cross section of agencies that employ clinical psychologists. For example, Sundberg's (1961) questionnaire was mailed to VA hospitals, state hospitals, institutions for the retarded, outpatient clinics, counselling centres, and university training clinics. The number of questionnaires sent out for the three surveys was 304, 551, and 249, respectively. The return rate for usable completed questionnaires was approximately 50 per cent.

Although many data were gathered, we will contrast the top ten tests as indicated in each of the surveys. Two types of scores are indicated for each test. TM refers to the total mention of the test. For example, of 251 respondents in the Lubin et al. (1971) survey, how many used a given test (e.g., the Rorschach)? WS refers to the weighted score rank—the percentage of the test's usage from, say, 251 respondents multiplied by the frequency of usage (i.e., 0-3 rating).

Among the top ten tests in Sundberg's (1961) surveys were four projectives, four intelligence tests, one objective personality test (MMPI), and one test to evaluate organicity (Bender-Gestalt). Interestingly, the composition of the top ten remained quite consistent in 1971 and 1976: five projectives, three intelligence tests, one objective personality test, and one test for organicity.

## Psychological Testing

However, over the fifteen years some definite changes did take place. The popularity of the Rorschach diminished in each successive survey, while that of the MMPI increased. Also, the Stanford-Binet declined substantially as the WISC (Wechsler Intelligence Scale for Children) and the WPPSI (Wechsler Preschool and Primary Scale of Intelligence) increased. (The TM and WS ranking of 1 for the WISC in the Brown and McGuire study undoubtedly points up sampling error, in that child treatment agencies were overrepresented.)

To summarise the trends, *objective* tests (IQ, personality, organicity) have gained in popularity and usage over the last two decades. However, the use of projective tests, although diminished, is still quite extensive.

Turning now to how testing is viewed in academia, let us consider two surveys designed to ascertain the attitude of academic psychologists toward projective testing. Thelen, Varble, and Johnson (1968) sent their questionnaire to representative faculty from seventy APA-approved clinical training programmes and received an 86 per cent return. In response to question 1: "Do you feel that knowledge and skill in the use of projective are as important as they used to be?" 75 per cent said yes, 11 per cent no, and 13 per cent were uncertain. In response to question 5: "Do you think that research generally supports the value of projective techniques?" 62 per cent said no, 12 per cent yes, and 22 per cent were uncertain. And in response to question 9: "Some of the major universities are cutting down on the semester hour time for teaching projective techniques," 51 per cent reacted favourably, 22 per cent unfavourably, and 25 per cent were neutral.

## PSYCHOMETRIC AND ETHICAL STANDARDS

In developing a good psychological test, three basic psychometric qualities must be maintained: norms, reliability, and validity. If they are missing, the extrinsic value of any given test can be seriously questioned. And once these values for a test are established, the practicing clinician must be fully aware of them. Given the delicate decisions that are often made on the basis of testing, as well as the emotionally charged social climate of testing today, the clinical psychologist would be doubly remiss in giving unreliable or invalid tests whose norms are not clearly defined.

Several booklets outlining technical recommendations for psychological tests have been published by the APA since 1954. The one of these, *Standards for Educational & Psychological Tests,* appeared in 1974. "Part of the stimulus for revision is an awakened concern about problems like invasion of privacy or discrimination against members of groups such as minorities or women.

Serious misuses of tests include, for example, labeling Spanish-speaking children as mentally retarded on the basis of scores on tests standardized on 'a representative sample of American children,' or using a test with a major loading on verbal comprehension without appropriate validation in an attempt to screen out large numbers of blacks from manipulative jobs requiring minimal communication". In short, the most recent revision of this manual is designed largely *to avoid the pitfalls of discriminatory use of tests.*

Let us now examine some of the recommendations in the latest APA manual. In discussing the various tests used by clinical psychologists later on, our appraisals of these tests will be based on how well they conform to the properties defined below.

## Norms

Norms refer to the range of scores obtained on a given test by the standardization sample. In practice, the standardization sample is both large and representative. To review, the bell-shaped distribution of scores is arranged so that there are three standard deviations above and three below the mean of all scores. Thus, at times norms may form standard deviation intervals above and below the mean. Sometimes norms are presented as percentile scores (e.g., a score at the mean is equal to the 50th percentile). The purpose of norms (which are usually presented at the beginning of the test manual) is to allow the tester to compare an individual score with the scores of other similar individuals. For example, how does a particular score on a college entrance examination compare with norms for successful college graduates?

Of course, there is always the danger that norms can be used in a discriminatory manner. Consider the following possibility. Suppose a poor high school student attains a score of 60 on an aptitude test. When comparing this score with norms for high school students of similar status, a score of 60 may fall in the 80th percentile. However, when comparing it with those of wealthy middle-class high school students, the same score of 60 may fall in the 40th percentile.

Clearly, several series of specific as well as general norms must be established for each test developed. Again, this is important not only for the test developer but also for the test user. Without appropriate norms, an individual's test score might be totally misinterpreted.' And as a result, the individual could suffer personal harm as a result of misclassification. This, unfortunately, has often happened when low IQ scores of foreign-speaking persons have been sweepingly interpreted as indicating "mental deficiency"; in fact, there were simply no appropriate norms for such people at the time. As the 1974 APA manual clearly states: "Norms presented in the test manual should refer to defined and clearly described populations. These populations should be the groups with whom users of the test will ordinarily wish to compare the persons tested".

## Reliability

A second important psychometric consideration in test construction is reliability. In general, *reliability* in a scientific sense means that a given observation or event can be reproduced at will under standard conditions. Reliability refers to whether a test score actually reflects the characteristic being measured (e.g., depression) or is simply an indication of chance factors. Such chance factors are often called "measurement error." Thus, if a test is described as reliable, the psychological examiner should be able to accept a given score as being a true indicator of the characteristic being assessed. Also, chance factors of measurement error should be at a minimum.

A test's reliability is expressed as a correlation coefficient ($r$ = 0.00 to 1.00). The closer it is to 1.00, the greater the reliability. By convention, however, a test whose reliability is equal to or greater than 0.80 is considered to be sufficiently reliable.

There are several ways to determine reliability. One of the easiest is to give the same test to the same individual (under similar conditions) on two separate occasions (i.e., test-retest). Then the two test scores are correlated. Although this is one of the methods most frequently used, it does have certain limitations. One of the problems is to insure that the test-retest interval is great enough so that practice or familiarity with the items do not affect the retest. On the other hand, if the interval is too long, educational, maturational, and other experiences may influence changed responses on the retest. An alternative approach is to develop parallel forms of the test. This, then, should mitigate the problems with the test-retest method. However, this is usually a time-consuming and costly procedure. Also, it is difficult to develop test items that have equal value and equal 'pull" for a given characteristic. This is especially true in projective testing.

A third method for determining test reliability is the "split-half" reliability approach. Here, instead of developing parallel forms, the existing test is divided into two (odd-numbered versus even-numbered items). The entire test is given as usual, but the total scores obtained on the basis of the odd— and even-numbered items are correlated. The problem here is that odd—and even-numbered items may not he carefully matched in the first place.

Clearly, then, even if a test manual reports respectable reliability levels, the careful examiner should be fully aware of how the reliability figure was determined and calculated. A mere statement that the test is "reliable" is obviously insufficient.

In addition to test reliability, there is the issue of inter-scorer reliability. In tests where scorer judgment may be at issue (e.g., responses to an IQ or projective test), one must be able to demonstrate that independent scorers, given equal training and operational definitions of correct and incorrect responses, are able to arrive at very similar conclusions (i.e., close to identical scores on a particular test). Here too a correlation of $r = 0.80$ would be considered a minima' acceptable criterion for inter-scorer reliability. In the absence of acceptable inter-scorer reliability, the usefulness of the test would indeed be questionable. In demonstrating good inter-scorer reliability, independent scores for both individual items and total test scores would be correlated.

## Validity

A high reliability coefficient, unfortunately, does not automatically ensure test validity. On the other hand, if the reliability coefficient is low, it is most unlikely that good validity will be attained. Let us now define validity. According to the APA manual, "Questions of validity are questions of what may be properly inferred from a test score; validity refers to the appropriateness of inferences from test scores or other forms of assessment. The many types of validity questions can, for convenience, be reduced to two: (a) What can be inferred about what is being measured by the test? (b) What can be inferred about other behaviour?" Put more simply, test validity indicates whether the test truly measures what it is supposed to measure. For example, in the case of a depression inventory, is the test really measuring depression? Are scores highly correlated with independent clinicians' ratings of depression? What is the correlation between scores on the depression inventory and other tests of depression? Do scores on the depression inventory relate to observed symptoms of depression (e.g., crying spells, sadness, poor sleep, loss of appetite, loss of energy)?

As with reliability, validity can be determined in many ways. Again, the mere statement that a test is valid or invalid is generally insufficient. The kind of validity that a test possesses is particularly important. Cronbach (1970) contends that "The question to ask is 'How valid is this test for the decision I wish to make' or 'How valid is the interpretation I propose for the test?'"

The first type of validity to be described, *face validity*, is not measured in a numerical or psychometric sense. Instead, it refers more to a global impression that the test appears to be reasonable that is, it *seems* to be measuring what it says it is measuring. The items appear to be related to the dimension (e.g.., anxiety, depression, schizophrenia) in question. However, as astutely pointed out by Cronbach (1970), many tests that appear to have face validity turn out to be very poor predictors of the dimension in question.

A more psychometrically related kind of validity is *content validity*. As the term implies, a test that has sufficient content validity is one in which ".....the behaviours demonstrated in testing constitute a representative sample of behaviours to be exhibited in a desired performance domain." For example, does a final examination ask questions about material actually covered in the course? To establish good content validity, the test developer usually takes the primary; dimension (e.g., anxiety or depression) and subdivides it into relevant subcategories. In this way, it becomes possible to construct items that are representative of all the subcategories. With this procedure, a truly representative sample of the universe of items emerges.

The next two types of validity—*criterion-related validity*—are extremely important. Here, the test score is correlated with a readily available external criterion. The first validity procedure is called *concurrent validity*. As the term implies, the test score is correlated with an external criterion that is *currently* available at the time of testing. An example of this might be the correlation of achievement test scores with current performance (i.e., grade-point average) in the classroom. By contrast, for *predictive validity*, the test score is correlated with an external criterion that *will* be available in the future. One example is the correlation of college entrance examination scores with subsequent success in college. In either instance (concurrent or predictive validity) the closer the correlation approaches $r = 1.00$, the greater the validity of the test.

Probably the most "amorphous" strategy for determining test validity is *construct validation*. It is amorphous in the sense that *construct validation* is a continuous process "based on an accumulation of research results" rather than being based on any one study. Therefore, construct validity on any given test is always a matter of interpretation. "In obtaining the information needed to establish construct validity, the investigator begins by formulating hypotheses about the characteristics of those who have high scores on the test in contrast to those who have low scores. Taken together, such hypotheses form at least a tentative theory about the nature of the construct the test is believed to be measuring".

An example of such a construct is anxiety. In researching the construct validity of a test of anxiety, the test developer has certain hypotheses as to how high-anxious and low-anxious people behave (e.g., high-anxious people have increased heart rates when placed in stressful situations). Thus, scores on an anxiety test may be correlated with subjects' heart rates in a behavioural stress situation. Also, the test may be correlated with other tests that presume to measure anxiety. Of course, the test is likely to be more valuable if the correlation with heart rate is higher than with the second test of anxiety. If the reverse were true, needless duplication of measurement would be represented by the two tests. Thus, in the ideal situation there should be moderate correlations between and among tests of anxiety and higher correlations with external criteria presumed to differentiate high-anxious and low-anxious individuals.

## Ethical Standards

As a professional serving the general public in the role of psychological examiner, the clinician is expected to act in an ethical manner. Although the importance of ethics has been acknowledged, we contend that this area receives too little attention during the training period. That is not to say that graduate clinical programmes avoid the issue. However, certain matters relating to ethical standards need 10 be underscored more carefully.

Probably the most crucial area is that of test selection. Although perhaps not typically seen in this light, the selection of tests with adequate norms, reliability, and validity for diagnostic purposes *is* the clinician's ethical responsibility. As previously noted, many critical decisions are often made on the basis of test scores and the ensuing psychological test report. In that light, let us consider the following medical analogy. 'What would happen if a physician made crucial medical judgments on the basis of diagnostic procedures of questionable reliability and validity? It is obvious that the examiner must thoroughly understand the research literature for the tests used. Although this seems so obvious as to be *given,* a recent survey suggests that many clinical psychologists pay more attention to their own convictions about a particular test than to the relevant research conclusions. In our judgment, this borders on the unethical.

A second ethical consideration relates to test security. This means that the examiner must safeguard the stimulus materials and the answers (e.g., on IQ and achievement tests) for many tests and inventories. Only those who administer, score, and interpret the tests should have access to such materials. 'Without test security, the validity of test scores would obviously be greatly reduced. Further, psychological testing is a serious endeavour; it does not and should never fall into the realm of "parlor games." A third major ethical responsibility relates to how test scores are sent to the referral party and what kind of information is given to the client.

For example, if the client scores 119 on the WAIS and the report is being forwarded to another clinical psychologist, one could normally assume that the referral party would know that this score is over one standard deviation above the mean. However, to report the score alone to a referral source who is not knowledgeable is neither sufficient nor appropriate. In this instance, it would be desirable to report that the WAIS IQ score of 119 means that the client is functioning in the upper end of the "bright-normal" range of intelligence and that the score falls in the 90th percentile. Also, it might be useful to explain what the 90th percentile means (i.e., the score is greater than that attained by 90 per cent of the population).

Equally important, the psychological examiner and/or the referring source should discuss the test results with the client. Too often the psychological

# Psychological Testing

examination has an unnecessary aura of mystery that probably impedes communication between tester and client as well as hindering performance. In the example of our 119 WAIS IQ score, direct feedback as to the actual score is not recommended.

However, feedback about the percentile equivalent and what it means and implies is quite consistent with the APA guidelines. Feedback about performance on "personality" tests should he handled the same way. Again, exact scores are neither recommended nor are they particularly useful to the client. But an overall description in layman's terminology should be given once all materials ire scored and interpreted.

We have highlighted only some of the primary ethical responsibilities of the psychological examiner. However, there are many more that become apparent in day-to-day clinical practice. The APA recommendations provide a more comprehensive overview of the issues.

## PSYCHOMETRIC TESTING

As mentioned previously, psychologists usually employ psychometric tests as an integral part of their assessments. However, psychologists rarely use psychometric tests or other standardised tools in isolation, and most of these tools require some degree of subjective interpretation. Indeed, psychologists are trained *not* to report psychometric test results in isolation from other pertinent clinical information, and will typically interpret test results in light of information gathered using other data-gathering methods (e.g. clinical interview, behavioural observations).

For example, it would be considered bad practice for a psychologist to report psychometric evidence about a client's intelligence without considering the influence of their educational and employment history, cultural background and medical history. According to Maloney and Ward (1976), the [u]ltimate meaningfulness and usefulness of test data are fundamentally dependent on a clinical consideration and integration of that data with other test data and non-test data by a person skilled in the process of assessment. Psychological assessment is a process and tests are tools that can be employed in the process.

What is psychometric testing? Psychometric testing is a method of assessment that is uniquely employed by psychologists, and the vast majority of psychometric tests should only be administered and interpreted by qualified psychologists. Gaudiness's (1995) survey of BPS psychologists found that 96 per cent of the sample employed psychological tests as a component of their forensic assessments. Psychometric testing involves the systematic measurement of individual differences along specified traits or dimensions. Hundreds of psychometric tests have been developed to measure a wide variety of psychological variables. Psychometric tests are designed to maximise the objectivity of assessment through standardization of their administration, scoring and interpretation. Standardization is intended to increase the reliability of measurement and ensure comparability of results across time and between and within individuals.

Standardization ensures that a test is always administered, scored and interpreted in the same way, regardless of who is taking or administering the test. A good psychometric test will have a test manual that describes test development, reliability and validity of the test, procedures for administration, scoring and interpretation and norms that allow for comparison of the subject's scores with population scores.

The latter allows the psychologist to determine how the subject performs relative to a comparison group with similar characteristics. Standardization of interpretation is ensured through the utilization of empirically derived norms which allow for the interpretation of individual scores.

Norms provide the basis for giving meaning to individual scores. They are used to evaluate an individual's performance in relation to a particular reference group. The individual taking the test needs to be sufficiently similar to the reference group for the norms to be appropriate.

For example, it would be problematic to evaluate a female's performance on a test against male norms. Psychometric tests differ according to the quality and detail of their normative samples. Some tests provide national norms whereas others only provide norms for specific subgroups (e.g. prisoners or psychiatric patients). Some tests provide different sets of norms (e.g. national norms and subpopulation norms) with which to compare an individual's scores, so the psychologist must be sure to select the most appropriate norms for comparison. For example, the Wechsler Adult Intelligence Scale has norms for different age groups, and the MCMI-III and MMPI-2 have norms for men and women. Inappropriate use of norms obviously undermines the validity of the test results and the expert opinion based on those results.

Measurement of psychological variables is far more problematic than in the physical sciences because psychological variables tend to be complex, difficult to define and intangible. This introduces scope for error even when standardised tests are employed. Standardised tests may still rely on subjective judgement to some extent, and psychologists may still disagree on the meaning of particular test scores or profiles.

Indeed, even the most highly regarded psychometric tests have "confidence intervals" for test scores that specify the probability that a score falls within specific parameters. For example, the Wechsler Adult Intelligence Scale provides such confidence intervals for IQ scores which means that the psychologist can say that they are "95 per cent certain that a person's score falls between *a* and *b*".

What makes a good psychometric test? A number of recommendations have been made about the criteria that should be met before psychometric tests are used to inform an expert opinion for court. Heilbrun (1992) recommends that a psychometric test should be based on sound theoretical principles and have solid empirical evidence regarding its utility. He states that there should be a test manual that describes test development (e.g. how norms were obtained), psychometric properties of the test and procedures for administration, scoring and interpretation. Psychometric properties should be scrutinised to ensure that the test has adequate reliability (i.e. consistency of measurement) and validity (i.e. whether it measures what it purports to measure) so as to justify its use in court.

A psychometric test should only be used if it is relevant to the legal issue or psychological construct underlying the legal issue, and should only be used for the purpose for which it was designed and validated. Pope et al (2002) warn psychologists against interpreting a test to fit a particular legal theory. Furthermore, psychometric tests should only be administered to individuals who are sufficiently similar to the normative group on which the test was standardised. A highly salient issue in South Africa is that the majority of psychometric tests have been developed and validated in Western cultures (usually North America and the UK), which raises questions about whether they can be validly applied cross-culturally to an African context, particularly where there may also be language barriers.

How should psychometric tests be used? Pope et al (2000) recommend that psychometric tests should only be employed if the psychologist has sufficient expertise and up-to-date training in its application. Psychologists will typically gain competence in the principles of psychometric testing and the use of certain psychometric instruments by virtue of their generic training. However, some psychometric tests require specialised training, and such training may sometimes be listed as a prerequisite in the test manual. Lack of the recommended training can be challenged in court and may undermine the credibility of evidence based on the test results. Pope et al (2000) also warn against the use of obsolete versions of tests, out-of-date norms, or modified forms of administration that have not been adequately researched.

Psychologists should adhere strictly to the standardised test procedures specified in the test manual. Pope et al (2000) recommend that psychologists should avoid the administration of psychometric tests without close and continuous monitoring (e.g. they should avoid asking the "client" to complete a self-report questionnaire at home).

The psychologist should always take note of the response style (e.g. faking good or faking bad; random responding) of the individual taking the test, as this may bias the results. Some psychometric tests have built-in measures of response style which help to determine whether the results may be invalid; these should always be reported.

Pope et al (2000) caution against making interpretations or inferences about scores on an invalid test profile and also recommend that psychologists note any factors that could undermine the validity of the test (e.g. visual problems, language barriers). Ackerman (1999) notes that it may be difficult to meet all the above criteria in every situation and recommends that the psychologist always be able to defend the use of the test whilst also acknowledging the possible effects of any shortcomings.

A more general issue relates to the release of raw data and test materials to non-psychologists. Ackerman (1999) points out that psychologists have an ethical and contractual duty to maintain the integrity of the psychometric tests they employ. Psychologists should not disclose raw test data, answer sheets or test questions to non-psychologists, including lawyers, for several reasons.

First, the raw data may be misused or misinterpreted by non-qualified individuals. If parties in a legal case request access to raw test data, then the psychologist should arrange for it to be sent in confidence to another psychologist acting on behalf of the party concerned. Second, test material is copyrighted and should not be copied or disclosed to individuals who have not purchased the test. Third, disclosure of raw data and test materials potentially places them in the public domain and this could compromise the validity of the test and its future use (Ackerman, 1999.

## POST-ASSESSMENT PHASE

### Report Writing

Psychologists will typically submit a written report which summarises the information gathered during the assessment and outlines their opinion. It is important to list all the sources of information used in the preparation of the report, highlight any important omissions and qualify opinions in light of any missing data. All data that are relevant to the legal issue should be reported. Psychologists should be aware that any notes that are related to the opinion are discoverable (i.e. can be obtained by lawyers or the court). The report will form the foundation for the expert evidence given in court, including cross-examination. The psychologist must therefore be prepared to justify every line in the report and should prepare thoroughly for any challenges that may arise during cross-examination. The psychologist must bear in mind that

# Psychological Testing

the other side may instruct another expert psychologist to testify or advise them in court about lines of cross-examination.

## Giving Evidence in Court

If an expert has followed the recommendations, then giving evidence in court should not be too daunting. Brodsky et al (2002) recommend that honesty, consistency and professional competency are the best policies when serving as an expert witness. Regardless of expertise and experience, the psychologist acting as an expert witness should prepare thoroughly and comprehensively before giving evidence in court. The contents of the report and the opinion should be discussed with the lawyer prior to giving evidence, and lines of questioning and potential areas for cross-examination should be discussed.

Psychologists should also familiarise themselves with the layout of the court room, court room procedures and court etiquette (e.g. taking the oath, proper forms of address).

## The Challenge of Maintaining Neutrality

One of the challenges of acting as an expert witness is to maintain objectivity and neutrality. Experts may feel pressurised to produce an opinion that is favourable to the side that has instructed them rather than one that is impartial. The extreme version of this is "the hired gun", who is instructed for the very reason that he or she will produce an "opinion" that is favourable to the side giving instruction. Such behaviour is not only unethical but also reduces the credibility of expert testimony in general. The pressure to produce a biased opinion may be subtle. For example, it is not unknown for lawyers to ask experts to change the opinions expressed in reports to suit their case.

Gaudiness's (1995) survey of BPS members, for instance, showed that 27 per cent of the 514 respondents had been asked to modify their reports, most commonly by solicitors (87%). For example, 32 psychologists reported that they had been asked alter their opinion, 19 reported being asked to remove reference to a document seen, and 44 reported being asked to remove unfavourable findings.

Compliance with these requests was reported by 22, 74 and 41 per cent of respondents respectively. Van Dorsten (2002) recommends that experts should constantly monitor their impartiality and view themselves as working for the court, regardless of who has instructed them. It is considered bad practice for an expert to change an opinion under pressure, or to omit information from a report, although it is considered acceptable to change a factual piece of information (preferably in an addendum).

One of the difficulties that can arise when a psychologist acts as an expert witness is confusion about their role. Psychologists are trained to adopt a therapeutic role and to act in the best interests of their clients. However, a psychologist acting as an expert witness is essentially working for the court, not the individual being assessed or the legal team that has instructed them. Sometimes, the roles of expert and "helping professional" can become confused, and this creates problems regarding the impartiality of the expert opinion that results.

## The Challenge of Sticking to the Expert Opinion

Another pressure that may be placed on expert witnesses is to express an opinion that goes beyond the psychological question they were instructed to address. This may involve offering a psychological opinion about an issue that was not directly assessed. Shapiro (2002) strongly recommends that experts adhere to only the areas dealt with in their report and never go beyond their data.

Experts may also be pressurised to address the "ultimate issue" (e.g. Is the person guilty of the crime? Should a mother have custody of her child? Should the plaintiff receive compensation for a psychological injury?). Although the expert may very well have a *personal* opinion about the ultimate issue, it is the responsibility of the Court to decide upon it. An expert should refuse to give an opinion about the ultimate issue on the grounds that it is not within their realm of expertise as a psychologist to make such a judgement.

## Risk Assessment

The concept of risk of harm to others plays an important part in both criminal and civil legal proceedings. Unlike psychometric testing, risk assessment is not an exclusive domain of psychologists. There is a long tradition whereby Courts have requested opinions regarding risk of harm to others from a range of professionals, including psychologists, psychiatrists, social workers and probation officers. This creates a social control function whereby the professional contributes to decision-making that has implications for both public safety and individual rights to freedom. For example, Criminal Courts may request an opinion regarding risk of future violence and/or sexual re-offending to inform decisions about the length of a prison sentence it should award, or in some countries, about issuing the death penalty. Family Courts may request an assessment of the risk of physical harm posed to children, which may result in a child being removed from the family.

The practice of risk assessment is controversial, particularly because empirical evidence regarding the reliability and validity (i.e. accuracy) of risk assessment is discouraging. This has led to a longstanding debate about

# Psychological Testing

whether mental health professionals (MHPs) should offer opinions about risk at all, especially in court cases, where the liberty of a person may be at stake.

A large body of research has been conducted over the years to provide MHPs with a legitimate scientific basis for risk assessment and to improve the reliability and validity of their assessments. The major scientific focus in the field has been on the search for valid risk factors for predicting violence and the development of standardised tools for measuring risk with a sufficient degree of accuracy. Research has focused on developing accurate methods for assessing risk in mentally disordered offenders (i.e. offenders with mental illness or personality disorders) and offenders who do not have a mental disorder. Despite considerable progress in the field, risk assessment still has major limitations. It is important for lawyers and psychologists to be aware of these limitations so that expert evidence about risk can be appropriately provided and adequately evaluated.

## Evidence About the Validity of Risk Assessment

The violence-risk assessment debate was sparked by the landmark 1966 case of Baxstrom v. Herald in the USA. The ruling in this case led to the release or transfer of 966 patients from maximum-security hospitals to the community or lower security institutions against the clinical judgement of MHPs. In the following four years, only 20 per cent of these released prisoners were reconvicted, the majority for non-violent offences. This raised serious questions about the ability of MHPs to accurately predict future violence, and several further studies in the USA provided similar evidence. This led to a debate about whether MHPs should be involved in risk assessment at all.

The evidence about imprecision was so compelling that the American Psychological Association (APA) (1978) stated that the validity of psychological predictions of violent behaviour...[is]...so poor that one could oppose their use on the strictly empirical grounds that psychologists are not competent to make such judgements. A similar conclusion was reached by the American Psychiatric Association (1974): Neither psychiatrists nor anyone else have reliably demonstrated an ability to predict future violence or 'dangerousness'. Neither has any special psychiatric 'expertise' in this area been established.

This position was strengthened by evidence that MHPs were not significantly better than lay-people at assessing risk of violence. Monahan (1981) reviewed the "first generation" of violence-risk assessment research from the 1960s and 1970s. He concluded that the best clinical research currently in existence indicates that psychiatrists and psychologists are accurate in no more than one out of three predictions of violent behaviour over a several-year period among institutionalized populations that had both committed violence in the past (and thus had a high base-rate for it) and who were diagnosed as mentally ill. In another review, Steadman (1980) concluded that [e]ven among what are generally considered extremely high risk groups, clinical estimations...[of violent behaviour]...rarely exceeded that which was obtainable simply by chance.

However, Monahan (1984) argued that methodological problems with the "first generation" research prevented conclusive statements about the validity of clinical predictions in all situations. Studies were criticised because predictive judgements were frequently old and inferred from MHPs' recommendations rather than from explicit judgements. Further, researchers used different definitions of violence and dangerousness across studies and followed up participants for different periods. In general, there was poor ability to generalise findings across populations, contexts and time-frames. There was a lull in validity studies between 1979 and 1993, followed by a "second generation" of violence-risk assessment research conducted in the early 1990s.

This led to a more optimistic outlook. Second generation studies suggested that MHPs had at least a modest ability to predict violence in hospital and community samples, and that their predictions were significantly more accurate than chance, especially amongst certain diagnostic groups (e.g. schizophrenics).

Otto (1992) reviewed the literature and concluded that "changing conceptions of dangerousness and advances in predictive techniques suggest that, rather than one in three predictions of long-term dangerousness being accurate, at least one in two short-term predictions are accurate". However, Otto cautioned that even under the best circumstances, MHPs would continue to make predictions where false positives were the most common type of error.

Borum (1996) argued that the apparent improvement in predictive accuracy in the "second generation" studies reflected an improvement in research methodology rather than real improvements in accuracy.

In particular, he noted that researchers had widened the criterion of violent behaviour beyond arrest or re-conviction rates to include self-reports and collateral reports of violent behaviour, thus reducing the rate of artefactual false-positives. McNiel and Binder (1995) also suggested that over-prediction may be exaggerated because MHPs are likely to make management/treatment decisions that bias outcomes in favour of their assessments. In other words, they are likely to subject individuals whom they consider to be a risk to others to preventive treatment or detention.

Risk assessment: Why is it so problematic? The problem of operationally defining violence and harm. A fundamental problem of violence-risk assessment relates to difficulties in defining violence *per se*. Different operational definitions employed by researchers and MHPs have implications for the reliability and validity of research findings and clinical judgements. According to Megargee, no definition of violence has proved completely successful. Although everyone knows what violence is, no-one has ever been able to define it adequately so that every possible instance of violent behaviour is clearly included within the definition while all the excluded behaviour is clearly non-violent.

So, different investigators who are all ostensibly studying the same phenomenon may in fact focus on quite different sorts of behaviour. According to De Zuluetta (1993), violence is an essentially human characteristic related to the meaning we give to destructive forms of interpersonal behaviour, the meaning being determined by social context.

The term "dangerousness" was used commonly in the risk assessment literature until the 1980s. Violence and dangerousness are related but not coterminous, and both are intrinsically related to the concept of risk. Floud and Young (1981) explain how notions of risk and harm (including that which flows from violent behaviour) relate to perceptions of danger, and how public opinion is formed regarding which social hazards, situations and acts constitute a danger.

The crux of their argument is that dangerousness is a matter of judgement or opinion, which necessarily has a subjective element.

"Danger", it is argued, reduces to a question of what people are prepared to tolerate and why, and not simply to a matter of what is in some degree damaging to them.

Floud and Young (1981) state that "dangers are unacceptable risks; we measure or assess the probability and severity of some harm and call it a risk; but we speak of danger when we judge the risk unacceptable". It is frequently asserted that danger is "in the eye of the beholder" and reflects social and personal values as much as objective estimates of anticipated harm.

The defining, labelling, and handling of dangerous behaviours and situations are very much influenced by the dominant values and power structures that exist in society. For example, social deviants labelled as 'mentally ill' have for several hundred years...aroused much societal apprehension and been major targets for preventive confinement...Yet many other categories of persons and groups who have quite glaringly demonstrated their dangerousness to society...do not seem to evoke similar concerns, nor are they readily subjected to indeterminate and preventive confinement in order to protect the community.

## The Base-Rate Problem

One of the reasons why violence is difficult to predict is because it is a rare event (i.e. it has a low base-rate), and rare events are inherently difficult to predict.

The general rule for any prediction system is that its error rate should be lower than the base-rate of the criterion being measured.

If this is not the case, more errors will occur from predicting the behaviour than from simply allowing events to occur by chance. So, when predicting low base-rate conditions such as serious violence, extremely accurate prediction tools are required in order to avoid a high error-rate.

Regrettably, violence-risk assessment tools (both clinical and actuarial) are relatively crude and imprecise, thus resulting in inaccuracies primarily in the direction of false-positives. High false-positive rates are a serious problem from an ethical point of view if sanctions (e.g. restrictions on liberty) are imposed upon all individuals considered to be at risk. There are also cost implications if scarce resources are channelled towards individuals who are not really in need. False-negatives generate a different set of costs.

They result in serious injury or loss of life in the worst outcome, either to members of the public or to the individual who has been incorrectly identified as "safe", as well as to the families of victims of violence. False-negatives also have consequences for MHPs who may face adverse social and political sanctions as a result. Some authors have criticised the use of false-positives and false-negatives to illustrate the invalidity of predictions of violence. According to Quen (1985), MHPs do not predict that violent behaviour *will* occur, but rather that an individual has a certain potential to act violently, and that there is a relative probability that this potential will be realised.

Whether this potential is realised is not indicative of the validity or invalidity of the prediction *per se* but on chance factors which the MHP cannot know in advance.

In other words, identifying false-positives as incorrect predictive judgements is misleading, because these false-positives may be just as potentially violent as the true-positives, but may simply not have been exposed to the chance factors which triggered violent behaviour in the true-positive group.

Towl and Crichton (1996) also argue that the uncertainty inherent in risk assessment is frequently overlooked, and that it is misleading to consider an initial risk assessment to be incorrect simply because something goes wrong.

They suggest that the validity of a risk assessment decision should be based upon an examination of the methodology used to inform the risk assessment process as well as the ethical basis of the decision-making. In relation to violence, Floud and Young (1981) assert that errors of prediction reflect "the fact that if we were to set them at liberty, [perhaps] only half of those we are at any time detaining as dangerous would do further harm...[However, this] does not mean that the other half are all in this sense innocent".

The problem with this position is that it makes the accuracy of prediction impossible to test. Any evidence that an individual has been falsely classified as potentially violent can be dismissed as luck; it is simply fortuitous that the person's potential for violent behaviour has not been triggered by chance factors. Those who make predictions can thus never be wrong.

## Decision-Making Biases

Some researchers have argued that over-prediction occurs due to specific features of decision-making and clinical judgement *per se*, such that violence prediction is prone to the same types of error that occur in other types of decisionmaking (e.g. a tendency to ignore base-rates).

This resulted in debate concerning the relative merits of clinical versus actuarial approaches to risk assessment. The interest in actuarial methods was influenced by Meehl's (1954) work.

He reviewed studies that demonstrated the superiority of actuarial prediction over clinical prediction in a range of settings. A substantial body of research has suggested that in almost all tasks, including the prediction of violence, actuarial formulas predict as well or better than clinical judgements. Consequently, a number of researchers have developed actuarial tools for assessing the risk of violence in specific populations.

Some of these tools are not strictly actuarial in that they combine purely actuarial data with clinical/professional expertise and thus can be better described as a form of structured clinical judgement.

The majority of risk assessment instruments have been developed and validated in North America so their applicability to the South African context is questionable. The best known examples include the Violence Risk Appraisal Guide, the Historical/Clinical/Risk Management 20, Psychopathy Checklist-Revised (PCL-R) and the Sex Offender Risk Appraisal Guide.

As with all psychological tests, the psychometric properties of any risk assessment instrument must be scrutinised before the instrument is used. The PCL-R has been identified as one of the best instruments for predicting violence and is incorporated into all of the instruments mentioned above.

Strong correlations have been found between PCL-R scores and violence (including sexual offending) amongst high-risk populations (e.g. prisoners and mentally disordered offenders). Salekin, Roger and Sewell (1996) conducted a meta-analysis of 18 studies that examined the association between psychopathy and violence.

They found a large effect size indicating the predictive validity of the construct of psychopathy. Some studies have identified psychopathy as a better predictor of violence than either psychiatric diagnosis or substance abuse. Similarly, the MacArthur Violence Risk Assessment Study identified psychopathy as the best predictor of violence out of the wide range of variables it examined.

However, despite its apparent superiority to other instruments, even the PCL-R has been heavily criticised on the grounds that it generates too high a high number of false positives to justify its use in the criminal justice system "where life and liberty decisions are at stake".

Despite their purported superiority to clinical judgement, actuarial tools still generate errors (chiefly false-positives) which some consider unacceptably high. Actuarial tools have also been criticised on the grounds that they can only demonstrate that an individual belongs to a group with a statistically higher than average risk of violence.

According to Morris and Miller (1985), the epistemology of actuarial or statistical prediction provides no grounds for the prediction of *individual* behaviour.

It refers by nature to predictions of the behaviour of defined *groups* of individuals. In other words, a statistical prediction does not *claim* to forecast the behaviour of any given individual, but rather the average behaviour of a group of like individuals.

A statistical prediction is a statement of a *condition* (i.e. membership in a class which possesses certain common attributes) and not the prediction of a *result* (i.e. future violent acts in each individual case). MHPs must still translate what group membership means for an individual and decide what action to take. Further, Gardner et al. (1996) argue that actuarial tools may be too costly, time-consuming or mathematically complicated to be of practical use in clinical settings.

The information required may not be available at the time of assessment, and some actuarial tools may not be applicable to the short-and medium-term predictions required for proper risk management. Maden (1998) criticised the emphasis on historical variables in many actuarial instruments on the basis that they tend to produce static estimates of risk. He argued that risk assessment should be seen as a dynamic process, where risk is constantly monitored, re-assessed, and coupled with *risk management*, which involves a series of decisions and interventions over time. It is now generally agreed that the best approach to risk assessment is to combine the best of both the clinical and actuarial approaches. However, there remains a lack of clarity about how best to achieve this systematically in order to arrive at probabilistic decisions about risk.

Borum (1996) argued that the emerging scientific knowledge about risk factors provides a framework for professional consensus and practice guidelines.

However, he noted the dearth of empirical guidance about how to systematically integrate relevant information to arrive at a decision about risk, and stated that "[t]he great challenge...is to integrate the almost separate worlds of research on violence prediction and the clinical practice of risk assessment. At present, the two domains rarely intersect".

There is also debate across disciplines about the extent to which scientific knowledge can inform and improve risk assessment practice. Towl and Crighton (1996) argue that whilst some aspects of the risk assessment process can benefit from a scientific approach, other aspects of the process are trans-scientific (i.e. they require solutions beyond the practical application of a scientific method).

They argue that a scientific approach can be employed to estimate the consequences and probability of a target behaviour occurring.

However, they make a distinction between the latter and the decision about what ought to be done in light of the prediction. They argue that decisions about what action should follow a prediction of violence are independent of the scientific method and are rooted in moral and ethical considerations about individual and collective rights and responsibilities. These decisions are based on value judgements about the appropriate balance that ought to be struck between the protection of society and the protection of individual rights to freedom (i.e. the extent to which society is willing to accept errors of prediction in order to prevent harm).

## The Majority of which were for Civil Forensic Assessment

The field of forensic psychology, or the application of psychological theories to issues in the law, has grown in the past twenty years since it became its own division (D41) within the American Psychological Association (APA). Initially, beginning around the early 1900's the forensic psychologist concentrated on the application of social psychology research to answer legal questions such as providing information to the court concerning the impact of sex discrimination on children in segregated schools, the scientific accuracy of classification of children in schools using intelligence tests, and psychological impact of forensic evidence such as eyewitness testimony or polygraph examinations.

Later, as psychologists began to develop specialities and proficiencies, the clinical forensic psychologist began to offer assistance to the courts when criminal defendants appeared to lack the requisite mens rea or state of mind due to mental illness.

Competency-to-proceed-to-trial as well as insanity cases now routinely utilize the clinical forensic psychologist's skills. Child custody disputes utilize the skills of the clinical forensic examiner to address the issue of what is in the best interests of the child. The forensic neuropsychologist addresses the psychological impact of brain damage from toxicity or other causes in personal injury lawsuits and the practitioner testifies to the community standard of care when licensing board complaints or malpractice actions are filed.

These and other legal issues provide the mainstay of the clinician who also practices in the legal arena.

One of the most important skills that the clinical psychologist has to offer to the courts is the ability to use standardized assessments to augment the clinical interview in evaluating the cognitions, emotions, and behaviour of the litigant.

However, the interpretation of the responses to these assessment tools is different when used in answering legal questions. This article will suggest issues to think about when applying clinical assessment instruments to a forensic population.

We also describe some of the forensic psychological assessment instruments that are now available in situations such as violence risk, competency, or impact from trauma. Finally, we emphasize the importance of integrating the results of the tests, clinical interview, and observations, together with the documents reviewed to place the results within the context of the current legal question and other behavioural descriptions.

## PSYCHOLOGICAL TESTS

There are several broad categories of psychological tests:

### IQ/Achievement Tests

IQ tests purport to be measures of intelligence, while achievement tests are measures of the use and level of development of use of the ability. IQ (or cognitive) tests and achievement tests are common norm-referenced tests. In these types of tests, a series of tasks is presented to the person being evaluated, and the person's responses are graded according to carefully prescribed guidelines.

After the test is completed, the results can be compiled and compared to the responses of a norm group, usually composed of people at the same age or grade level as the person being evaluated. IQ tests which contain a series of tasks typically divide the tasks into verbal (relying on the use of language) and performance, or non-verbal (relying on eye–hand types of tasks, or use of symbols or objects). Examples of verbal IQ test tasks are vocabulary and information (answering general knowledge questions). Non-verbal examples are timed completion of puzzles (object assembly) and identifying images which fit a pattern (matrix reasoning). IQ tests (e.g., WAIS-

IV, WISC-IV, Cattell Culture Fair III, Woodcock-Johnson Tests of Cognitive Abilities-III, Stanford-Binet Intelligence Scales V) and academic achievement tests (e.g. WIAT, WRAT, Woodcock-Johnson Tests of Achievement-III) are designed to be administered to either an individual (by a trained evaluator) or to a group of people (paper and pencil tests). The individually-administered tests tend to be more comprehensive, more reliable, more valid and generally to have better psychometric characteristics than group-administered tests. However, individually administered tests are more expensive to administer because of the need for a trained administrator (psychologist, school psychologist, or psychometrician).

## Attitude Tests

Attitude test assess an individual's feelings about an event, person, or object. Attitude scales are used in marketing to determine individual (and group) preferences for brands, or items. Typically attitude tests use either a Thurston Scale, or Likert Scale to measure specific items.

## Neuropsychological Tests

These tests consist of specifically designed tasks used to measure a psychological function known to be linked to a particular brain structure or pathway. They are typically used to assess impairment after an injury or illness known to affect neurocognitive functioning, or when used in research, to contrast neuropsychological abilities across experimental groups.

## Personality Tests

Psychological measures of personality are often described as either objective tests or projective tests. The terms "objective test" and "projective test" have recently come under criticism in the Journal of Personality Assessment. The more descriptive "rating scale or self-report measures" and "free response measures" are suggested, rather than the terms "objective tests" and "projective tests," respectively.

## Objective Tests (Rating Scale or Self-report Measure)

Objective tests have a restricted response format, such as allowing for true or false answers or rating using an ordinal scale. Prominent examples of objective personality tests include the Minnesota Multiphasic Personality Inventory, Millon Clinical Multiaxial Inventory-III, Child Behaviour Checklist, Symptom Checklist 90 and the Beck Depression Inventory. Objective personality tests can be designed for use in business for potential employees, such as the NEO-PI, the 16PF, and the OPQ (Occupational Personality Questionnaire), all of which are based on the Big Five taxonomy.

The Big Five, or Five Factor Model of normal personality, has gained acceptance since the early 1990s when some influential meta-analyses found consistent relationships between the Big Five personality factors and important criterion variables. Another personality test based upon the Five Factor Model is the Five Factor Personality Inventory – Children (FFPI-C.).

## Projective Tests

Projective tests allow for a freer type of response. An example of this would be the Rorschach test, in which a person states what each of ten ink blots might be. Projective testing became a growth industry in the first half of the 1900s, with doubts about the theoretical assumptions behind projective testing arising in the second half of the 1900s. Some projective tests are used less often today because they are more time consuming to administer and because the reliability and validity are controversial. As improved sampling and statistical methods developed, much controversy regarding the utility and validity of projective testing has occurred.

The use of clinical judgement rather than norms and statistics to evaluate people's characteristics has convinced many that projectives are deficient and unreliable (results are too dissimilar each time a test is given to the same person). However, many practitioners continue to rely on projective testing, and some testing experts (e.g., Cohen,Anastasi) suggest that these measures can be useful in developing therapeutic rapport. They may also be useful in creating inferences to follow-up with other methods. The most widely used scoring system for the Rorschach is the Exner system of scoring. Another common projective test is the Thematic Apperception Test (TAT), which is often scored with Westen's Social Cognition and Object Relations Scales and Phebe Cramer's Defense Mechanisms Manual. Both "rating scale" and "free response" measures are used in contemporary clinical practice, with a trend toward the former. Other projective tests include the House-Tree-Person Test, Robert's Apperception Test, and the Attachment Projective.

## Sexological Tests

The number of tests specifically meant for the field of sexology is quite limited. The field of sexology provides different psychological evaluation devices in order to examine the various aspects of the discomfort, problem or dysfunction, regardless of whether they are individual or relational ones.

## Direct Observation Tests

Although most psychological tests are "rating scale" or "free response" measures, psychological assessment may also involve the observation of people as they complete activities. This type of assessment is usually conducted with families in a laboratory, home or with children in a classroom.

The purpose may be clinical, such as to establish a pre-intervention baseline of a child's hyperactive or aggressive classroom behaviours or to observe the nature of a parent-child interaction in order to understand a relational disorder. Direct observation procedures are also used in research, for example to study the relationship between intrapsychic variables and specific target behaviours, or to explore sequences of behavioural interaction.

The Parent-Child Interaction Assessment-II (PCIA) is an example of a direct observation procedure that is used with school-age children and parents. The parents and children are video recorded playing at a make-believe zoo. The Parent-Child Early Relational Assessment is used to study parents and young children and involves a feeding and a puzzle task. The MacArthur Story Stem Battery (MSSB) is used to elicit narratives from children.

The Dyadic Parent-Child Interaction Coding System-II tracks the extent to which children follow the commands of parents and *vice versa* and is well suited to the study of children with Oppositional Defiant Disorders and their parents.

## Test Security

Many psychological tests are generally not available to the public, but rather, have restrictions both from publishers of the tests and from psychology licensing boards that prevent the disclosure of the tests themselves and information about the interpretation of the results. Test publishers consider both copyright and matters of professional ethics to be involved in protecting the secrecy of their tests, and they sell tests only to people who have proved their educational and professional qualifications to the test maker's satisfaction. Purchasers are legally bound from giving test answers or the tests themselves out to the public unless permitted under the test maker's standard conditions for administration of the tests.

# 6

# Cognitive Assessment

The assessment of cognitive functioning in forensic situations is similar to its assessment in clinical, school, and neuropsychological settings. The gold standard is the Wechsler Adult Intelligence Scale, Third Edition (WAIS-III). In some situations that do not require precise measurement a standard clinical interview is used to estimate that a person is functioning within the normal range of cognitive abilities.

However, in many forensic situations accurate assessment of cognitive abilities may make an important difference in the outcome of the case. This is particularly true if a criminal defendant is charged with a serious crime such as in the case of a 15 year old who is waived into adult court for killing someone or if a mentally retarded person is facing the death penalty. In civil personal injury cases, the WAIS-III results that indicate a large difference between verbal and performance IQs might corroborate the damage from an accident and suggest further neuropsychological testing and rehabilitation that is necessary.

## Clinical Forensic Psychology Assessment Instruments

Cognitive Areas:
- Weschler Adult Intelligence Scale-Third Edition (WAIS-III)
- Weschler Abbreviated Intelligence Scale (WASI)
- Test of Non Verbal Intelligence (TONI)
- Weschler Achievement Tests (WIAT)
- Wide Range Achievement Tests (WRAT-IV)
- Repeatable Battery for Assessment of Neuropsychological Symptoms (RBANS)
- Halstead-Reitan Battery
- Luria-Nebraska Battery.

Personality & Emotional Domains:
- Minnesota Multiphasic Personality Inventory (MMPI-2)

## Cognitive Assessment

- Minnesota Multiphasic Personality Inventory for Adolescents (MMPI-A)
- Personality Assessment Inventory (PAI)
- Rorschach Technique.

Trauma Impact:
- Trauma Symptom Inventory (TSI)
- Trauma Symptom Checklist for Children (TSCC)
- Detailed Assessment of Post-traumatic Stress (DAPS).

Violence Risk Assessment:
- Violence Risk Assessment Guide (VRAG)
- Sexual Offender Risk Assessment Guide (SORAG)
- Historical Clinical Risk (HCR-20)
- Psychopathy Check List (PCL-R).

Competency:
- Function of Rights in Interrogation (Grisso)
- MacArthur Competency Assessment Tool for Criminal Adjudication (MACCAT-CA)

Malingering:
- Structured Interview of Reported Symptoms (SIRS)
- Miller Forensic Assessment of Symptoms (MFAST)
- Test of Memory Malingering (TOMM)
- Validity Indicator Profile (VIP).

The Wechsler Abbreviated Scale of Intelligence (WASI) calls for the administration of four of the subtest, two in the verbal and two in the performance areas and is a better test to use than choosing several WAIS-III subtests, if time is limited and cognitive functioning is not a primary concern in the legal questions to be answered. However, not administering all of the WAIS-III subtests makes it impossible to discuss working memory and other index scores that may be even more important in forensic cases questioning cognitive abilities than in clinical cases.

Obviously, interpretation of this and other cognitive tests must consider culture and language background as well as impact of emotional functioning on the person's cognitive abilities. The Test of Non Verbal Intelligence (TONI) may be substituted for the WAIS-III in those whose knowledge of English is poor. Forensic examiners use tests that can assess for reading level [Wide Range Achievement Tests (WRAT), Wechsler Achievement Tests (WIAT)] and other areas of academic achievement together with the WAIS-III that can help determine competency to enter into contracts or other legal issues requiring cognitive ability or what is often called, 'knowingly' in the law. Neuropsychological tests such as the Weschler Memory Scale (WMS),

Repeatable Battery for Neuropsychological Screening (RBANS), and parts of the Halstead-Reitan or Luria Nebraska Batteries are also used by the practitioner although the actual batteries require neuropsychological training.

## Personality Assessment

The best practice in forensic assessment of emotional or personality functioning is to use both objective and subjective assessment instruments. Subjective assessment utilizes many of the behavioural assessment inventories as well as the Rorschach test using the Exner scoring system. The traditional clinical observational techniques of the psychologist are often subjected to cross-examination for reliability and validity issues under the Frye and Daubert admissibility standards so structured assessment inventories are more acceptable in court testimony. The objective tests include the Minnesota Multiphasic Personality Inventory (MMPI-2) and the Personality Assessment Inventory (PAI). The latter is being used more frequently in courts as more psychologists have been trained in its interpretation. While there are many other objective tests published, we will discuss these as they remain the most popular ones offered to and accepted by the court.

## MMPI-2 & MMPI-A

We have already noted, with the MMPI-2, some of the situational or contextual constraints necessary to properly interpret findings in a forensic context. Nevertheless, provided one does not rely blindly on these results, the MMPI-2 has several scales that make it helpful in a forensic context. The traditional validity scales, coupled with more recent developments such as consistency and stability scales make it useful when one is trying to determine the likelihood that a particular configuration reflects current stressors, pre existing stressors, and whether or not the current mental state is likely to persist over time. New scales, such as the 'Fake Bad Scale' consisting of mostly physical complaints, have proven useful in adjudicating personal injury cases.

The FP scale may be of some assistance in determining malingering of extremely psychotic symptoms. NCS has programs for various forensic contexts (criminal, personal injury, child custody) which may be of some assistance, but do not provide normative data for these different types of assessments. The Caldwell Report generates a list of Forensic Critical Items, consisting of items which forensic psychologists have frequently noted reflect situational as opposed to pathological interpretations. The items are provided as an interview guide for the clinician.

The MMPI-A is constructed of similar items that are for use with adolescents. This can be of assistance when evaluating someone under the age of 18 as the MMPI-2 norms do not cover this population. Given the need to assess the emotional stability of adolescents, this test is gaining in popularity in the courts.

## PAI

The Personality Assessment Inventory, is relatively recent compared to the MMPI, but it shows great promise in forensic contexts. Its use of a Likert scale gives the examinee a more finely tuned set of a responses than the true and false on the MMPI. In addition, it has a number of scales that are helpful in forensic work that the MMPI does not. It has a number of scales that measure complex PTSD from trauma, a dimension not adequately assessed by the MMPI-2. Its validity scales are also configured differently, and are potentially more useful in forensic work. Finally, the PAI does have actual norms based on correctional populations which the MMPI-2 does not.

## Rorschach

The Rorschach Technique, although denounced by some critics because of its subjectivity, can in fact be very valuable in forensic contexts, especially when using the more objective Exner administrative and scoring system. For instance, the Special Scores provide information on the formal structure of a thought disturbance such as schizophrenia. An individual trying to malinger a mental illness can certainly fake the content of such an illness (i.e. say he or she is hearing voices), but is less likely to fake the formal structure of a thought disorder that would be reflected in the Special Scores. Therefore, a discrepancy between claimed symptoms and the absence of a formal thought disturbance can alert the examiner to the possibility of malingering. The scores that Exner refers to as D and Adjusted D may be very helpful in determining, both in civil and in criminal cases, the temporal stability of a poor stress tolerance. In criminal cases, is it just reflecting the effects of incarceration, or does it also suggest that the individual had such difficulties around the time of an offense? In personal injury cases, is the incapacity to tolerate stress, both recent and situational, reflecting possibly the effects of a recent trauma, or was it more chronic, suggesting the impact of a pre-existing condition?

In addition to these specific areas, the Rorschach assists the examiner to understand how the person perceptually organizes his or her world when there is little to no structure. For example, does she or he see one small detail that determines the final perception or does she or he use the entire area of the card to form concepts. The standardization of card presentation in the Exner system helps clarify some of the former inconsistencies noted by critics of the test. The Rorschach's subjectivity may actually complement the findings on the WAIS-III especially if someone has what we call 'street smarts' rather than formal intelligence or comes from a culture where the WAIS-III norms have not been tested. And, it offers a nice balance to the more formal questions found on the standardized objective tests to better understand the complexity of emotional functioning in an individual.

## Assessment of Malingering

As noted above, the assessment of malingering is critical in forensic work. Many assessment instruments have been developed and validated within the past ten to fifteen years. They assess malingering both of psychosis and of cognitive deficits. We give these as examples.

### STRUCTURED INTERVIEW OF REPORTED SYMPTOMS (SIRS)

Perhaps the best known of these instruments for assessing malingering of psychiatric symptoms is the Structured Interview of Reported Symptoms (SIRS), developed by Rogers. Rogers essentially took the major areas of malingering, and developed scales based on each. He has, for example, scales based on improbable and absurd symptoms, unusual symptom combinations, exceedingly rare symptoms, symptoms that occur suddenly, symptoms that are overly specified, symptoms that are exaggerated, and symptoms that are inconsistently described over time

### Test of Memory Malingering (TOMM)

The Test of Memory Malingering (TOMM) may be very helpful in looking at the malingering of cognitive impairment. The TOMM provides norms of honest responders who have traumatic brain injury or dementia and compares those to people deliberately trying to do poorly.

### Validity Indicator Profile (VIP)

The Validity Indicator Profile (VIP) is a more complex test that compares an individual's performance on tests of varying degrees of difficulty, both verbal and non-verbal, to individuals who are honest responders, deliberately doing poorly, or are careless, or distractible. Both the VIP and the TOMM reported above, are based on symptom validity testing, a forced choice situation where, by chance alone, an examinee will get 50% of the items correct.

## Assessment of Trauma

The assessment of the impact from trauma has become an important tool for the forensic psychologist given the numbers of cases involving physical, sexual and psychological maltreatment of children and violence against women. There are a number of tests now available. We only discuss the following here:

- Trauma Symptom Inventory (TSI),

## Cognitive Assessment

- Trauma Symptom Checklist for Children (TSCC), & Trauma Symptom Checklist for Youth (TSCY).

Perhaps the most widely used standardized test to measure the impact from interpersonal trauma is the self administered Trauma Symptom Inventory (TSI) developed by Briere. Used as a research tool by those who assessed for trauma over the past twenty years, it has been published as a clinical and forensic tool with norms based on clinical and forensic populations.

The adult version, the TSI, lists 100 symptoms using a likert scale and the results are shown in a graph similar to the MMPI-2 with the mean set at 50 and the standard score at 10. There are two validity and reliability scales and ten clinical scales that correspond to those psychological problems known to impact trauma victims such as anxious arousal, depression, irritability, intrusive experiences, dissociation and sexual concerns. The computerized scoring version has three summary scales that can assist in separating out recent and chronic abuse problems. One major limitation to using this test is the person's language skills and reading level, especially comprehension of the items.

There are three versions of these trauma tests that are useful when child abuse is alleged. Two are directly administered to children, one with (TSCC) and one without (TSCC-A) questions about impact from sexual abuse.

Although geared to a third grade reading level, it may be necessary to read the symptoms to the child and help him or her respond to the proper four-point likert scale items. A new test that uses parents' responses to children's symptoms (TSCY), especially those younger than eight years old which is the cut off range for the TSCC, has just appeared in the marketplace. Given the need to assess the reliability and validity of children's abuse claims, these tests may be helpful in child maltreatment, dependency, and custody cases.

## Detailed Assessment of Post-traumatic Stress (DAPS)

The Detailed Assessment of Post-traumatic Stress (DAPS) is another standardized test developed by John Briere to assess for the level of PTSD in those who have been exposed to trauma.

It is a more complicated test to administer as first the various types of trauma exposure must be ascertained, the one for which the person has current complaints chosen, and then questions are answered based on the selected traumatic event. Results have reliability and validity indexes as well as pre and post trauma responses discussed and shown on a graph.

## Specific Forensic Assessment Tests

These tests are developed specifically to measure a specific psycholegal issue unlike the other tests that have other clinical uses in addition to assessing for forensic problems. Developers of these tests note that many courts have become disenchanted with traditional clinical assessment because it does not deal directly with legal issues.

For instance, a personality test may tell us that someone is psychotic, but it does not directly assess the issue of a person's mental state at the time of an offense.

An intelligence test may tell us that a person is mentally retarded but it does not tell us whether that individual was competent to waive their Miranda Rights or whether he or she is competent to stand trial. Forensic assessment instruments try to do just that: they take a legal construct and try to operationalize it.

## COMPETENCY ASSESSMENT

### Function of Rights in Interrogation

Grisso has a series of instruments, called the Function of Rights in Interrogation, designed to assess whether a juvenile or an adult defendant was competent to waive Miranda rights. The instrument consists of four parts. The first deals with the defendant stating in his or her own words what each legal right means.

The second part of the test asks the defendant to compare each of the statements to other legally relevant statements and see whether they are the same or different. The defendant is then asked to define each term, and finally to use their understanding in a series of scenarios dealing with right to silence, assistance of counsel, and behaviour in a courtroom.

### MacArthur Competency Assessment Tool – Criminal Adjudication

The MACCAT-CA is an instrument for the assessment of competency to stand trial. It divides competency into knowledge, appreciation, and reasoning, giving the defendant several scenarios from which they have to show their ability to reason it through and discuss it coherently with their attorney. It has the distinct advantage, over other briefer competency assessments, in that it actually assesses ability to assist counsel, which other instruments do not.

### Criminal Responsibility Assessment Scales

The Rogers Criminal Responsibility Assessment Scales are essentially a complex coding system that takes all of the material from psychological assessments, clinical interviews, third party information, and collateral records and contacts, puts them into a decision tree that assists in a determination of criminal responsibility.

## VIOLENCE RISK ASSESSMENT

### MacArthur Studies

The MacArthur studies of violence was a fifteen year project that looked at existing studies about the risk of predicting violence and analysed both the strengths and deficiencies in them. They identified over thirty risk assessment domains that needed to be accounted for when providing a risk assessment of an individual. They further broke down these domains into psychological, sociological, contextual, demographic, and biological areas. Using these domains, new risk assessment instruments have begun to appear in the marketplace.

### Historical Clinical and Risk Factors (HCR-20)

The HCR-20 provides a structured interview format based on current available research to aid the clinician in inquiring into areas that have been found important in current research on violence risk assessment. This is one of the few assessment instruments in this category that measures both static or unchangeable demographic factors together with changes from clinical treatment.

### Actuarial Assessment

Another approach to assessment of violence has been the use of actuarial assessments. While psychologists maintain that we cannot predict whether violence will occur, we can assign an assessment of risk based on actuarial tables. Most of these actuarial assessments are based on values given to fixed items that have been found to be associated with people who continue to commit violent acts.

Thus, a person who has committed one violent act at age 16 might be more likely than one who committed his or her first violent act at age 50 to reoffend. Using complicated mathematical formulas it is possible to compare one individual using these variables to others who have similar characteristics. Psychologists are most likely to use these actuarial instruments when asked for an opinion on someone's risk for further violence, such as when someone is seeking an early parole from prison, or for predatory sex offenders who are being considered for release or involuntary hospitalization in a forensic setting.

### Violence Risk Assessment Guide (VRAG)

One of the most popular violence risk actuarial instruments is called the VRAG. It is based on static variables such as age, sex, family composition etc. that can be applied to an individual to determine his or her risk of future violence. It has been formed on people who have been incarcerated. It can give the likelihood of recidivism over seven, ten, and fifteen year periods. A major limitation is that it does not take into account any dynamic variables such as response to treatment or situational life changes that might change someone's risk of using violence.

## Sexual Offender Risk Assessment Guide

The Sex Offender Risk Assessment Guide (SORAG) is an actuarial similar to the VRAG that gives the same type of risk assessment of an individual's likelihood of reoffending sexually. It has some different variables that have been found to be important in differentiating the predatory sex offender from the violent criminal.

## Psychopathy Check List – Revised (PCL-R)

PCL-R is a structured interview that Hare identified as defining psychopathy which he divided into two factors: anti-social behaviour and interpersonal style. The psychopath has fascinated people throughout history because of his (rarely but sometimes, her), lack of empathy for others and no remorse when engaging in hurtful behaviour. Although the concept fell out of favour for awhile, research continued to point to a biological basis for the psychopath's behaviour, requiring forensic psychological examination.

The PCL-R requires significant training in proper administration and scoring but may provide the forensic examiner with a detailed understanding of a particular type of criminal and is highly correlated with the potential for violent behaviour.

## PSYCHOLOGICAL TESTS

A psychological test is an instrument designed to measure unobserved constructs, also known as latent variables. Psychological tests are typically, but not necessarily, a series of tasks or problems that the respondent has to solve. Psychological tests can strongly resemble questionnaires, which are also designed to measure unobserved constructs, but differ in that psychological tests ask for a respondent's maximum performance whereas a questionnaire asks for the respondent's typical performance.

A useful psychological test must be both valid (i.e., there is evidence to support the specified interpretation of the test results) and reliable (i.e., internally consistent or give consistent results over time, across raters, etc.). It is important that people who are equal on the measured construct also have an equal probability of answering the test items correctly. For example, an item on a mathematics test could be "In a soccer match two players get a red card; how many players are left in the end?"; however, this item also requires

# Cognitive Assessment

knowledge of soccer to be answered correctly, not just mathematical ability. Group membership can also influence the chance of correctly answering items (differential item functioning). Often tests are constructed for a specific population, and this should be taken into account when administering tests. If a test is invariant to some group difference (e.g. gender) in one population (e.g. England) it does not automatically mean that it is also invariant in another population (e.g. Japan).

## Psychological Assessment

Psychological assessment is similar to psychological testing but usually involves a more comprehensive assessment of the individual. Psychological assessment is a process tha involves the integration of information from multiple sources, such as tests of normal and abnormal personality, tests of ability or intelligence, tests of interests or attitudes, as well as information from personal interviews.

Collateral information is also collected about personal, occupational, or medical history, such as from records or from interviews with parents, spouses, teachers, or previous therapists or physicians. A *psychological test* is one of the sources of data used within the process of assessment; usually more than one test is used.

Many psychologists do some level of assessment when providing services to clients or patients, and may use for example, simple checklists to assess some traits or symptoms, but psychological assessment is a more complex, detailed, in-depth process. Typical types of focus for psychological assessment are to provide a diagnosis for treatment settings; to assess a particular area of functioning or disability often for school settings; to help select type of treatment or to assess treatment outcomes; to help courts decide issues such as child custody or competency to stand trial; or to help assess job applicants or employees and provide career development counseling or training.

## Interpreting Scores

Psychological tests, like many measurements of human characteristics, can be interpreted in a *norm-referenced* or *criterion-referenced* manner. Norms are statistical representations of a population. A norm-referenced score interpretation compares an individual's results on the test with the statistical representation of the population. In practice, rather than testing a population, a representative sample or group is tested. This provides a group norm or set of norms. One representation of norms is the Bell curve (also called "normal curve").

Norms are available for standardized psychological tests, allowing for an understanding of how an individual's scores compare with the group norms. Norm referenced scores are typically reported on the standard score (z) scale or a rescaling of it. A criterion-referenced interpretation of a test score compares an individual's performance to some criterion other than performance of other individuals.

For example, the generic school test typically provides a score in reference to a subject domain; a student might score 80% on a geography test. Criterion-referenced score interpretations are generally more applicable to achievement tests rather than psychological tests. Often, test scores can be interpreted in both ways; a score of 80% on a geography test could place a student at the 84th percentile, or a standard score of 1.0 or even 2.0.

# 7
# Psychology of Criminal Behaviour

*Rational Choice Theory:* Dr. William Glasser, MD coined the term choice theory. According to many criminologists, choice theory is perhaps the most common reason why criminals do the things they do. This theory suggests that the offender is completely rational when making the decision to commit a crime (Siegel, 2005). The variety of reasons in which one offends can be based on a variety of personal needs, including: greed, revenge, need, anger, lust, jealousy, thrills, and vanity. The rational choice theory has its root in the classical school of criminology which was developed by Italian "social-thinker" Cesare Beccaria. Classical criminology suggests that "people have free will to choose criminal or conventional behaviors…and that crime can be controlled only by the fear of criminal sanctions (Siegel).

Inside the rational choice theory there are three models of criminal behaviour: rational actor, predestined actor, and victimized actor. The rational actor proposed that individuals choose whether to commit a crime. With this belief, crime could simply be controlled by increasing the penalty of offending (Burke, 2001). The predestined actor proposes that criminals cannot control their personal urges and environment, thus, inducing them to commit crime. The way to solve this problem would then be to change the biological, sociological, and psychological environment of the offender. Finally, the victimized actor model proposes that crime is the result of the offender being a victim of an unequal society. Thus, the crime could be controlled by reforming legislation (Burke, 2001).

## DEFINITION AND MEASUREMENT OF CRIMINAL BEHAVIOUR

To fully understand the nature of how genes and the environment influence criminal behaviour, one must first know how criminal behaviour is defined. Law in our society is defined by social and legal institutions, not in biology

(Morley & Hall, 2003). Therefore determining what constitutes criminal behaviour can envelope a wide variety of activities and for that reason, researchers tend to focus on the wider context of antisocial behaviour. Authors Morley and Hall (2003), who have investigated the genetic influences on criminal behaviour, point out three different ways to define antisocial behaviour. First is equating it with criminality and delinquency, which both involve engaging in criminal acts.

Criminality can lead to arrest, conviction, or incarceration for adults, while delinquency is related to juveniles committing unlawful acts (Rhee & Waldman, 2002). Information can be collected using court and criminal records, as well as self report surveys to analyse the influences that were present. Secondly, they advise individuals to define antisocial behaviour is through criteria used to diagnose certain personality disorders. More specifically, they mean those personality disorders, such as Antisocial Personality Disorder, which is associated with an increased risk in criminal activity. A final measure suggested for defining antisocial behaviour is by examining personality traits that may be influential in the criminal behaviour of individuals. Traits such as aggressiveness and impulsivity are two traits that have been investigated the most (Morley & Hall, 2003). Further details of disorders and personality traits associated with criminal behaviour will be discussed later in the paper.

With regards to determining the effects the environment plays in criminal behaviour there are fewer resources available. Observational studies and reports submitted by parents are two sources, but not everyone agrees on the validity of information collected from these sources. Three additional sources that most researchers cite when gathering information about both genetic and environmental influences are twin, family, and adoption studies (Tehrani & Mednick, 2000).

## Personality Disorders and Traits

Personality traits and disorders have recently become essential in the diagnosis of individuals with antisocial or criminal behaviour. These traits and disorders do not first become evident when an individual is an adult, rather these can be seen in children. For that reason it seems logical to discuss those personality disorders that first appear in childhood. Attention Deficit Hyperactivity Disorder (ADHD), Conduct Disorder (CD), and Oppositional Defiance Disorder (ODD) are three of the more prominent disorders that have been shown to have a relationship with later adult behaviour (Holmes, Slaughter, & Kashani, 2001).

ODD is characterized by argumentativeness, noncompliance, and irritability, which can be found in early childhood (Holmes et al., 2001). When a child with ODD grows older, the characteristics of their behaviour also change and more often for the worse. They start to lie and steal, engage in vandalism, substance abuse, and show aggression towards peers (Holmes et al., 2001). Frequently ODD is the first disorder that is identified in children and if sustained can lead to the diagnosis of CD (Morley & Hall, 2003). It is important to note however that not all children who are diagnosed with ODD will develop CD.

ADHD is associated with hyperactivity-impulsivity and the inability to keep attention focused on one thing (Morley & Hall, 2003). Holmes et al. (2001) state that, "impulse control dysfunction and the presence of hyperactivity and inattention are the most highly related predisposing factors for presentation of antisocial behaviour". They also point to the fact that children diagnosed with ADHD have the inability to analyse and anticipate consequences or learn from their past behaviour. Children with this disorder are at risk of developing ODD and CD, unless the child is only diagnosed with Attention Deficit Disorder (ADD), in which case their chances of developing ODD or CD are limited. The future for some children is made worse when ADHD and CD are co-occurring because they will be more likely to continue their antisocial tendencies into adulthood (Holmes et al., 2001).

Conduct Disorder is characterized with an individual's violation of societal rules and norms (Morley & Hall, 2003). As the tendencies or behaviours of those children who are diagnosed with ODD or ADHD worsen and become more prevalent, the next logical diagnosis is CD. What is even more significant is the fact that ODD, ADHD, and CD are risk factors for developing Antisocial Personality Disorder (ASPD). This disorder can only be diagnosed when an individual is over the age of eighteen and at which point an individual shows persistent disregard for the rights of others (Morley & Hall, 2003). ASPD has been shown to be associated with an increased risk of criminal activity. Therefore, it is of great importance that these early childhood disorders are correctly diagnosed and effectively treated to prevent future problems.

Another critical aspect that must be examined regarding antisocial or criminal behaviour is the personality characteristics of individuals. Two of the most cited personality traits that can be shown to have an association with antisocial or criminal behaviour are impulsivity and aggression (Morley & Hall, 2003). According to the article written by Holmes et al. (2001), antisocial behaviour between the ages of nine and fifteen can be correlated strongly with impulsivity and that aggression in early childhood can predict antisocial acts and delinquency. One statistic shows that between seventy and ninety percent of violent offenders had been highly aggressive as young children (Holmes et al., 2001). These personality traits have, in some research, been shown to be heritable.

## Neurochemicals in Criminal and Anti-Social Behaviour

Neurochemicals are responsible for the activation of behavioural patterns and tendencies in specific areas of the brain (Elliot, 2000). As seen in the Brunner et al. study, there have been attempts to determine the role of neurochemicals in influencing criminal or antisocial behaviour. Included in the list of neurochemicals already cited by researchers are monoamine oxidase (MOA), epinephrine, norepinephrine, serotonin, and dopamine.

Monoamine oxidase (MAO) is an enzyme that has been shown to be related to antisocial behaviour. Specifically, low MAO activity results in disinhibition which can lead to impulsivity and aggression (Elliot, 2000). The Brunner et al. study is the only one to report findings of a relationship between a point mutation in the structural gene for MAOA and aggression, which makes the findings rare. However, there has been other evidence that points to the conclusion that deficiencies in MAOA activity may be more common and as a result may predispose individuals to antisocial or aggressive behaviour (Brunner et al., 1993). MAO is associated with many of the neurochemicals that already have a link to antisocial or criminal behaviour. Norepinephrine, serotonin, and dopamine are metabolized by both MAOA and MAOB (Elliot, 2000). While, according to Eysenck (1996), MAO is related to norepinephrine, epinephrine, and dopamine, which are all related to the personality factor of psychosis.

Serotonin is a neurochemical that plays an important role in the personality traits of depression, anxiety, and bipolar disorder (Larsen & Buss, 2005). It is also involved with brain development and a disorder in this system could lead to an increase in aggressiveness and impulsivity (Morley & Hall, 2003). As Lowenstein (2003) states, "studies point to serotonin as one of the most important central neuro-transmitters underlying the modulation of impulsive aggression". Low levels of serotonin have been found to be associated with impulsive behaviour and emotional aggression. In addition, children who suffer from conduct disorder (which will be discussed later), have also been shown to have low blood serotonin (Elliot, 2000). Needless to say, there is a great deal of evidence that shows serotonin is related to aggression, which can be further associated with antisocial or criminal behaviour. Dopamine is a neurotransmitter in the brain that is associated with pleasure and is also one of the neuro transmitters that is chiefly associated with aggression. Activation of both affective (emotionally driven) and predatory aggression is accomplished by dopamine (Elliot, 2000). Genes in the dopaminergic pathway have also been found to be involved with Attention Deficit Hyperactivity Disorder (ADHD) (Morley & Hall, 2003). In one study cited by Morley and Hall (2003), a relationship was found between the genes in the dopaminergic

## Psychology of Criminal Behaviour

pathway, impulsivity, ADHD, and violent offenders. Obviously, from this list of neurochemicals it seems plausible that there is a genetic component to antisocial or criminal behaviour.

### Twin, Adoption, and Family Studies

There has been great debate between researchers regarding the outcomes of twin, adoption, and family studies. Some claim that these studies support the notion of a genetic basis to criminal behaviour (Tehrani & Mednick, 2000). On the other hand, some have concluded that there is not enough evidence from these twin, family, and adoption studies to profess that genetics do play a role in antisocial or criminal behaviour (Lowenstein, 2003). To understand why there are such conflicting opinions, one must first look at the available studies that have been conducted.

Twin studies are conducted on the basis of comparing monozygotic (MZ) or identical twins and their rates of criminal behaviour with the rates of criminal behaviour of dizygotic (DZ) or fraternal twins. Ordinarily these studies are used to assess the roles of genetic and environmental influences. If the outcomes of these twin studies show that there is a higher concordance rate for MZ twins than for DZ twins in criminal behaviour, then it can be assumed that there is a genetic influence (Tehrani & Mednick, 2000). A study conducted looked at thirty two MZ twins reared apart, who had been adopted by a non-relative a short time after birth. The results showed that for both childhood and adult antisocial behaviour, there was a high degree of heritability involved (Joseph, 2001). This study was of particular importance because it examined the factor of separate environments.

Another researcher studied eighty-five MZ and one hundred and forty-seven DZ pairs and found that there was a higher concordance rate for the MZ pairs. Ten years later after checking police records of these same twins, two other researchers concluded that there was a fifty-four percent heritability of liability to crime (Joseph, 2001). Around the same time of the study just mentioned, two researchers studied forty-nine MZ and eighty-nine DZ pairs, but found no difference in the concordance rates. They concluded therefore that in respect to common crime, hereditary factors are of little significance (Joseph, 2001). Many other twin studies have been conducted, but there is concern over the validity of those studies and their ability to separate out the nature and nurture aspects; therefore other sources of information should be examined.

Adoption studies are critical in examining the relationship that exists between adopted children and both their biological and adoptive parents because they assume to separate nature and nurture. Studies have been conducted that test for the criminal behaviour of the adopted-away children, if their biological parents had also been involved with criminal activity. In Iowa, the first adoption study was conducted that looked at the genetics of criminal behaviour. The researchers found that as compared to the control group, the adopted individuals, which were born to incarcerated female offenders, had a higher rate of criminal convictions as adults.

Therefore this evidence supports the existence of a heritable component to antisocial or criminal behaviour (Tehrani & Mednick, 2000). Another study in Sweden also showed that if a biological background existed for criminality, then there was an increased risk of criminal behaviour in the adopted children. In Denmark, one of the largest studies of adopted children was conducted and found similar results to the previous studies. The defining feature of the Denmark study was that the researchers found a biological component for criminal acts against property, but not for violent crimes (Joseph, 2001). Children whose biological fathers had been convicted of property crimes were more likely to engage in similar behaviour, when compared to those biological fathers who had been convicted of violent crimes.

According to an article by Jay Joseph (2001), who studied all of the minor and major adoption studies, the majority of researchers have found and agreed upon the non-significance of genes in violent crime. This reestablishes the findings from the studies mentioned already in that there may be a genetic component to antisocial behaviour or that genes influence criminal behaviour, but specifically for property offenses. Family studies are the third type of instrument used to assess the relationship between genetics and environmental influences on criminal or antisocial behaviour. Research in this field has probably been the least accepted by psychologists and other scholars because of the degree of difficulty in separating out nature and nurture in the family environment. Children experience both the influence of their parents' genes and also the environment in which they are raised, so it is difficult to assign which behaviours were influenced by the two factors. Twin studies have this flaw, as stated earlier, but it is more prevalent in family studies. An additional concern with family studies is the inability to replicate the results, therefore leading to a small number of studies. Regardless of these drawbacks, one family study in particular should be acknowledged for its findings.

Brunner, Nelen, Breakefield, Ropers, and van Oost (1993) conducted a study utilizing a large Dutch family. In their study they found a point mutation in the structural gene for monoamine oxidase A (MAOA), a neurochemical in the brain, which they associated with aggressive criminal behaviour among a number of males in that family (Alper, 1995). These males were reported to have selective MAOA deficiency, which can lead to decreased concentrations of-hydroxyindole-3-acetic acid (5-HIAA) in cerebrospinal fluid. Evidence suggests that low concentrations of-HIAA can be associated with impulsive aggression. These results have not been confirmed in any additional family studies, which lead to a need for more studies to determine if other families share similar results (Brunner et al., 1993). However, this one family study does seem to suggest that genetics play an important role in antisocial or criminal behaviour.

## Environmental Influences

Thus far it has been established through research and various studies that genetics do influence criminal or antisocial behaviour. Researchers agree on the point that genes influence personality traits and disorders, such as the ones just mentioned. However, researchers also agree that there is an environmental component that needs to be examined. Environmental influences such as family and peers will be discussed, as well as a look into the social learning theory.

The family environment is critical to the upbringing of a child and if problems exist then the child is most likely to suffer the consequences. We have seen the problems associated with a child who is diagnosed with ADHD and how that can influence antisocial or criminal behaviour. In relation to that, some researchers have claimed that it is the family environment that influences the hyperactivity of children (Schmitz, 2003).

The researchers in this article specifically identify family risk factors as poverty, education, parenting practices, and family structure. Prior research on the relationship between family environment and child behaviour characterizes a child's well being with a positive and caring parent-child relationship, a stimulating home environment, and consistent disciplinary techniques (Schmitz, 2003). Families with poor communication and weak family bonds have been shown to have a correlation with children's development of aggressive/criminal behaviour.

Therefore it seems obvious to conclude that those families who are less financially sound, perhaps have more children, and who are unable to consistently punish their children will have a greater likelihood of promoting an environment that will influence antisocial or delinquent behaviour. Another indicator of future antisocial or criminal behaviour is that of abuse or neglect in childhood. A statistic shows that children are at a fifty percent greater risk of engaging in criminal acts, if they were neglected or abused (Holmes et al., 2001). This has been one of the most popular arguments as to why children develop antisocial or delinquent behaviours.

One additional research finding in the debate between genetic and environmental influences on antisocial or criminal behaviour has to deal with the age of the individual. Research seems consistent in recognizing that heritability influences adult behaviour more than environmental influences, but that for children and adolescents the environment is the most significant factor influencing their behaviour (Rhee & Waldman, 2002). As an adult, we have the ability to choose the environment in which to live and this will either positively or negatively reinforce our personality traits, such as aggressiveness. However, children and adolescents are limited to the extent of choosing an environment, which accounts for the greater influence of environmental factors in childhood behaviours. Another significant factor in the development of antisocial or delinquent behaviour in adolescence is peer groups. Garnefski and Okma (1996) state that there is a correlation between the involvement in an antisocial or delinquent peer group and problem behaviour. One of the primary causes as to why this occurs can be traced back to aggressive behaviour in young children. When children are in preschool and show aggressive tendencies towards their peers, they will likely be deemed as an outcast. This creates poor peer relationships and relegates those children to be with others who share similar behaviours. A relationship like this would most likely continue into adolescence and maybe even further into adulthood. The similar tendencies of these individuals create an environment in which they influence one another and push the problem towards criminal or violent behaviour.

Social learning theory has been cited as way to explain how the environment can influence a child's behaviour. Using this theory to explain the aggressive or antisocial behaviour of a child means that a child observes aggressive behaviour between parents, siblings, or both. As a result, the children believes that this aggressive behaviour is normal and can therefore use it themselves because they do not see the harm in acting similar to their parents (Miles & Carey, 1997). As stated earlier, interaction between family members and disciplinary techniques are influential in creating antisocial behaviour. Using the social learning theory these two factors are also critical in the development of aggression.

Children who are raised in an aggressive family environment would most likely be susceptible to experiencing a lack of parental monitoring, permissiveness or inconsistency in punishment, parental rejection and aggression. The exposure to such high levels of aggression and other environmental factors greatly influences and reinforces a child's behaviour. A significant point that should be known however is the fact that other research has supported the notion that genetics do influence levels of aggression, which stands in opposition to the social learning theory (Miles & Carey, 1997).

## Gene-Environment Interactions

There are theories, however, concerning genetic and environmental influences, which seem to suggest an interaction between the two and one such theory is the general arousal theory of criminality. Personality psychologist

Eysenck created a model based on three factors known as psychoticism, extraversion, and neuroticism, or what is referred to as the PEN model (Eysenck, 1996). Psychoticism was associated with the traits of aggressive, impersonal, impulsive, cold, antisocial, and un-empathetic. Extraversion was correlated with the traits of sociable, lively, active, sensation-seeking, carefree, dominant, and assertive.

Finally, neuroticism was associated with anxious, depressed, low self-esteem, irrational, moody, emotional, and tense (Eysenck, 1996). Through research and surveys, Eysenck found that these three factors could be used as predictors of criminal behaviour. He believed this to be especially true of the psychoticism factor and that measuring it could predict the difference between criminals and non-criminals. Extraversion was a better predictor for young individuals, while neuroticism was a better predictor for older individuals (Eysenck, 1996). An important point about these factors and the personality traits associated with them is that most of them have already been found to be heritable (Miles & Carey, 1997). Understanding Eysenck's original model is critical to assessing the general arousal theory of criminality, which suggests an interaction between factors. Research has shown that criminality is strongly correlated with low arousal levels in the brain. Characteristics related to low arousal levels include lack of interest, sleepiness, lack of attention, and loss of vigilance. Eysenck (1996) believed that these characteristics were similar to the personality factor of extraversion. Individuals with low arousal levels and those who are extraverts need to seek out stimulation because they do not have enough already in their brains.

Therefore, the premise of the general arousal theory of criminality is that individuals inherit a nervous system that is unresponsive to low levels of stimulation and as a consequence, these individuals have to seek out the proper stimulation to increase their arousal. Under this theory, the proper stimulation includes high-risk activities associated with antisocial behaviour, which consists of sexual promiscuity, substance abuse, and crime (Miles & Carey, 1997). A significant fact that must be pointed out though is that not every individual with low arousal levels or those who are extraverts will seek those high risk activities just mentioned. It takes the right environment and personality to create an individual with antisocial or criminal tendencies and that is why this theory can be considered to take into account both factors of genetic and environmental influences.

## CRIMINAL BEHAVIOUR AND PERSONALITY DISORDERS

In addition to the research showing that the gene responsible for production of monoamine oxidase has a possible link to criminality, some evidence has also shown a possible link between other genes. One area of personality research in molecular genetics that has received a lot of attention is the trait of novelty-seeking, and novelty seeking is a personality trait often associated with criminality. Research has indicated that the single DRD4 gene may account for 10% of the genetic variance in relation to novelty-seeking (Sloan, 2000). This conclusion is highly controversial because in subsequent studies there has been both replication of the original findings, and failure to replicate in other studies. It seems most molecular genetic approaches in relating heritability of personality characteristics such as criminality to a single gene frequently suffer from failures in replication.

Some studies have demonstrated a genetic link between ADHD, CD, and ODD and criminality. However, there are possible alternate explanations for a greater rate of criminality for those who have suffered from these disorders that the paper failed to mention. It has been shown that people evoke certain responses from their environment. It is plausible that children suffering from these disorders are treated in a different manner than normal children due to the responses that they evoke, and it is because of these environmental differences that they are more prone to criminal behaviour. Say a child suffering from ADHD is having problems in school, they may be placed in a remedial class in which there is a greater rate of delinquency. This would be a very important environment difference that could contribute greatly to future criminality. Other children may also socialize less with children with these disorders, which could plausible lead to anti-social behaviour.

While it is possible that in some cases the relationship between these disorders and criminality is not direct byproduct of genes, but rather as a byproduct of the same environment. There have been studies on ADHD in relation to a multitude of environmental factors, including everything from nutrition to environmental toxins.

For example, a studies have been done that indicate an increased time spent viewing television in children was related to a decreased attention span and ADHD. There have also been studies showing a relationship between television viewing and desensitization to violence, which could influence criminal behaviour. I am not proposing that it is watching TV that is the major factor in these disorders, or in criminality, I am just trying to illustrate that perhaps there is some environmental factor that could influence criminality as well as disorders such as ADHD.

In addition to ADHD, CD, and ODD, other disorders have shown to influence criminality as well. Studies have shown that there is a higher occurrence of disorders such as schizophrenia, major depressive disorder, bipolar disorder, just to name a few. It is possible that having these personality disorders gives one a greater predisposition toward criminal behaviour. A

## Psychology of Criminal Behaviour

Swedish study found that the occurrence of major mental disorders in prisoners to be%, as well as a 20% occurrence of personality disorders (Rasmussen, 1999). Other studies have given different values for the occurrence, but in most cases the research agrees that there is a much higher incidence of these mental disorders in those who commit crimes.

Genetics has shown to be a major factor in the occurrence of many of these disorders. There have been studies that examine the rate of personality disorders such a schizophrenia, psychosis, and manic? depressive illness in adopted children. We can hypothesize that if adopted children are more likely to suffer from such disorders if their biological parents are or were afflicted, it would be indicative of a genetic basis for the disorder. Research done by Leonard Heston in 1960 examined children of schizophrenic mothers that were removed after birth and raised by foster parents. Out of a total of 47 children examined, Heston found that nine of them were diagnosed with sociopathic personalities and antisocial behaviour, and four of the 47 children developed schizophrenia. Heston also found behavioural abnormalities in many of the other children (Eysenk, 1982). This study shows a significant increase in the rate of personality disorders in the progeny of an affected parent, in comparison to population statistics on these mood disorders. In fact, when both parents are affected by a personality disorder the rate of occurrence in their offspring is even higher.

Not only do adoption studies support a genetic basis for personality disorders that are shown to have a relation to criminality, but twin studies as well. Statistics show a high concordance between identical and non-identical twins for schizophrenia and manic depression. Analysis of the statistics clearly show the genetic basis for these disorders: For schizophrenia the concordance in identical twins was 60%, compared to only 10% in non-identical twins, and the normal frequency being 1% in northern European populations. Similarly, manic depression showed a 70% concordance between identical twins, a 15% concordance between non-identical twins, and again only a 1% frequency in the normal population (Russo & Cove, 1995). This research supports the theory that genetics play a crucial role in these personality disorders.

## PERSONALITY DISORDERS AND TRAITS

Personality traits and disorders have recently become essential in the diagnosis of individuals with antisocial or criminal behaviour. These traits and disorders do not first become evident when an individual is an adult, rather these can be seen in children. For that reason it seems logical to discuss those personality disorders that first appear in childhood. Attention Deficit Hyperactivity Disorder (ADHD), Conduct Disorder (CD), and Oppositional

Defiance Disorder (ODD) are three of the more prominent disorders that have been shown to have a relationship with later adult behaviour (Holmes, Slaughter, & Kashani, 2001).

ODD is characterized by argumentativeness, noncompliance, and irritability, which can be found in early childhood (Holmes et al., 2001). When a child with ODD grows older, the characteristics of their behaviour also change and more often for the worse. They start to lie and steal, engage in vandalism, substance abuse, and show aggression towards peers (Holmes et al., 2001). Frequently ODD is the first disorder that is identified in children and if sustained can lead to the diagnosis of CD (Morley & Hall, 2003). It is important to note however that not all children who are diagnosed with ODD will develop CD.

ADHD is associated with hyperactivity-impulsivity and the inability to keep attention focused on one thing (Morley & Hall, 2003). Holmes et al. (2001) state that, "impulse control dysfunction and the presence of hyperactivity and inattention are the most highly related predisposing factors for presentation of antisocial behaviour" (p.184).

They also point to the fact that children diagnosed with ADHD have the inability to analyze and anticipate consequences or learn from their past behaviour. Children with this disorder are at risk of developing ODD and CD, unless the child is only diagnosed with Attention Deficit Disorder (ADD), in which case their chances of developing ODD or CD are limited. The future for some children is made worse when ADHD and CD are co-occurring because they will be more likely to continue their antisocial tendencies into adulthood (Holmes et al., 2001).

Conduct Disorder is characterized with an individual's violation of societal rules and norms (Morley & Hall, 2003). As the tendencies or behaviours of those children who are diagnosed with ODD or ADHD worsen and become more prevalent, the next logical diagnosis is CD. What is even more significant is the fact that ODD, ADHD, and CD are risk factors for developing Antisocial Personality Disorder (ASPD). This disorder can only be diagnosed when an individual is over the age of eighteen and at which point an individual shows persistent disregard for the rights of others (Morley & Hall, 2003). ASPD has been shown to be associated with an increased risk of criminal activity. Therefore, it is of great importance that these early childhood disorders are correctly diagnosed and effectively treated to prevent future problems.

Another critical aspect that must be examined regarding antisocial or criminal behaviour is the personality characteristics of individuals. Two of the most cited personality traits that can be shown to have an association with antisocial or criminal behaviour are impulsivity and aggression (Morley & Hall, 2003). According to the article written by Holmes et al. (2001), antisocial

*Psychology of Criminal Behaviour*

behaviour between the ages of nine and fifteen can be correlated strongly with impulsivity and that aggression in early childhood can predict antisocial acts and delinquency. One statistic shows that between seventy and ninety percent of violent offenders had been highly aggressive as young children (Holmes et al., 2001). These personality traits have, in some research, been shown to be heritable.

## Criminality Is a Product of Genes and Environment

In considering the roles of genetics and environment on criminal behaviour, or any behaviour for that matter, I think the best explanation is that there is a complex interaction between one's inherited traits and the environment in which he or she lives. Although the idea of environmental influences seems rather intuitive, regardless of knowledge regarding heredity and biological factors, it is surprising that some may have considered criminal behaviour to be solely a result of genetics. I propose that the debate of nature versus nurture now is not whether genetics or environment influence behaviour, but how complex the interaction between these factors is.

Despite the relative lack of reliability and validity in twin, adoption, and family studies, they still provide valuable insight into the roles of heredity and environment in criminal behaviour. However, it seems that most studies of this kind focus on the role of heredity in influencing behaviour. It would be interesting to see whether any studies with adopted children have examined the role of environment in criminal behaviour. Most adoption studies examine the correlation between criminality in the biological parents of adopted children, but what about the correlation between the children and their adopted parents who are crucial to their environment?

The influence of neurochemicals on criminal and antisocial behaviour are indicative of a genetic component to such behaviours. However, a more complete explanation of neurochemical influences is that they reflect the complex interactions between genetics and environment. There is evidence that the expression of genes is influenced by a wide variety of environmental factors. Therefore, it is very possible that disorders relating to such chemicals as serotonin and dopamine could be caused by stressful environmental situations. If environment affects the regulation of gene expression and, in turn, the activity of neuro transmitters that modulate behaviour, this kind of interaction may be a significant factor in the development of criminal and antisocial behaviour. Environmental and genetic factors influence antisocial and criminal behaviour in childhood versus adulthood seems somewhat incomplete. While it is true that adults have more control of their environment than children, I do not think that children are necessarily affected more by environment and adults are influenced more by heredity. Inherited traits provide the foundation by which people are able to learn and respond to their environment. An adult's personality is the combination of traits and learned behaviour patterns that have been established throughout childhood. Thus, although it is true that adults have more control over their current environment, They are still heavily influenced by both their current environment and by past exposure to environmental factors.

The social learning theory is a good way to explain the influence of environment on antisocial behaviour in children, and does not necessarily have to oppose the notion of genetic influence on behaviour as well. Rather, it should be considered part of a larger theory or model that could describe how environment and genetics interact. Eysenck's general arousal theory, which suggests such an interaction, could be modified to encompass the social learning theory, providing a more complete model to explain how upbringing and inherited traits interact to influence criminal behaviour.

Genetics and environmental factors are so intertwined, that it seems impossible to separate them in explaining how people are caused to engage in criminal acts. It is important for society as a whole to take responsibility in preventing the advent of criminal and antisocial behaviour in children via programs to provide children with healthy, enriching environments. A eugenic approach to preventing antisocial behaviour is immoral and impinges on human rights, but taking an active approach to ensure positive environmental influences would be appropriate.

## PSYCHOLOGICAL AND OSYCHIATRIC THEORIES OF CRIMINAL BEHAVIOUR

Psychological and psychiatric theories of criminal behaviour emphasise individual propensities and characteristics in explanations of criminality. Whether the emphasis is on conditioned behaviour, the development of parental attachment, or the psychoanalytic structure of the human personality, these approaches see the wellsprings of human motivation, desire, and behavioural choice as being firmly rooted in the personality.

Some of the earliest positivists were convinced that criminal behaviour was a result of genetic abnormality. Lombroso advanced the notion of atavism, which stated criminals represented a savage, earlier form of humankind. Lombroso compiled a list of physical features that were associated with criminals, which included protruding eyes, long arms, tattoos and large jaws. He tested convicts and those who had 5 or more of these atavisms were deemed to be born criminals. However his research was based on male criminals in a Sicilian jail which was inadequate as a control group as it had an overrepresentation of Sicilians who naturally demonstrated several of

## Psychology of Criminal Behaviour

Lombrosos atavisms. Sheldons somatic typology theory was based on a large sample of males in rehabilitation institutions. He listed three major somatypes (or body types); endomorphs: obese, soft, and rounded people who were fun loving and sociable; mesomorphs: muscular, athletic people who were assertive, vigorous, and bold; ectomorphs: tall, thin, with a well developed brain who were introverted, sensitive, and nervous.

Sheldon thought that mesomorphs were most likely to become criminals. However unrepresentative samples were used and he may have been confusing causation with correlation – just because there is an association between body type and deviant behaviour doesn't mean that the body type/biology caused the deviant behaviour. Also neither of these theories take into account female crime. Some criminologists believe that criminal behaviour is genetic. There are two types of studies which try to draw the link between inherited traits and criminal behaviour:

1) twin studies which compare the criminal behaviour between identical (monozygotic) twins and fraternal (dizygotic) twins. These studies have found that there is a greater similarity of criminal behaviour between identical twins than between fraternal twins.

    As identical twins are genetically identical, and because of their similarity of criminal behaviour, it is suspected that criminal behaviour is genetically linked.

    DiLalla and Gorresman (1990) did a metanalysis of 4 decades of twin research into criminality and concluded that the average concordance rate for fraternal twins was 22% and for identical twins, 51%. The influence of heredity is higher for property crime than it is for violent crime.

2) adoption studies which look at the criminal behaviour of children that have been adopted. This allows for the separation of biological effects on criminal behaviour from the environmental effects on criminal behaviour. If the behaviour of both the adopted child and the biological parent behaviour is criminal, then there is support for a genetic basis for crime. Studies conducted in Europe show that the criminality of a biological father is a good predictor of criminality in an adopted offspring. The likelihood of criminal behaviour is greatest in an adopted child if both the biological and adoptive parents are criminal. Adoption Study of Cloninger et al (1982) examined children whose biological parents were criminals. The crime rate for children was 4 times greater if biological parents were criminals, 2 times greater if adopted parents were criminals (12% and 6%, respectively).

Neurotransmitters are chemicals that allow for the transmission of electrical impulses in the brain and are the brain\'s way of processing information. They are not normally involved with the regulatory nervous system (as hormones are, although some glands trigger neurotransmitters, or neuropeptides, as well as hormones), but with the Central Nervous System and higher-order cognitive functioning. As such, they have become of great interest to criminologists who study things like antisocial personality and psychopathy which are believed to manifest brain systems with neurotransmitter levels "out of balance". It is well documented that alcoholism and drug dependence are associated with differences in neurotransmitter levels. In fact, the biggest research problem in studying neurotransmitters is finding criminal research subjects who aren't already "self-medicated" on alcohol or drugs. Although a person's normal neurotransmitter levels are determined genetically at birth, it is quite easy to manipulate them with drugs (medications for the mentally ill, stop-smoking pills), with diet (sugar, caffeine, chocolate, food additives), with stress (stressful environmental conditions), and with altitude (hypoxia is a condition mimicking the effect of neurotransmitter imbalance at altitudes above 3,800 feet above sea level).

There are not that many neurotransmitters within the central nervous system, and the three most commonly studied ones are serotonin, dopamine, and norepinephrine. Antisocial people have significantly lower levels of serotonin than ordinary people do. Schizophrenics have significantly lower levels of dopamine, and cocaine addicts have higher levels of dopamine. Levels of norepinephrine have also been associated with antisocial behaviour. Serotonin is probably the most important neurotransmitter in criminology. As stated previously, antisocial people have lower levels of serotonin. This may be either genetic or environmental, because neurotransmitter balances are constantly changing as memories are stored in the brain. Every new memory permanently changes the neural pathway structure, thus creating the opportunities for neurotransmitter imbalances. People who are genetically endowed with lower serotonin levels ("born antisocials") may therefore "grow out of it", and likewise, someone who is born with normal serotonin levels may develop an antisocial personality (what is called "reduced serotonergic activity" or a "serotonin uptake problem"). Reduced serotonic activity and crime is one of the strongest connections in biopsychological criminology.

Brain wave activity has been studied, and the general finding is that criminals have slower brain waves, i.e., slower EEG activity. Whether this is an indicator of a central or autonomic nervous system problem depends upon the researcher. The work of criminologist Hans Eysenck points at one of the reasons for why criminals can beat the lie detector is because their slower autonomic nervous system results in their not being easily stimulated, hence they seek out exciting, criminal behaviour in a "stimulus hunger". Mednick, a sociobiologist, points out that criminals have a lower rate of skin conductance response (SCR), the time it takes the skin to conduct electrical current. He argues that this affects the ability of criminals to benefit from negative reinforcement, and since fear is the most powerful reinforcer known to psychology, criminals experience no fear or anxiety. Lobe dysfunction, which can occur with head injury or birth trauma, also has been studied in criminals. Prisoners often report having had a head injury involving loss of consciousness, and 80% of violent criminals had greater than average birth complications. There's evidence to suggest that frontal lobe dysfunction may characterise violent offenders while temporal lobe dysfunction may characterise sex offenders. Research involving newer imaging techniques e.g. MRI and CT is ongoing.

Freud never really had much to say about crime, other than it was most likely motivated by guilt, committed by people with overdeveloped superegos, and characterised by unconscious errors (Freudian slips) which appeared to represent a desire to get caught and be punished. The inconsistencies in this (is why its often said that there is no purely Freudian theory of crime other than the idea the criminals want to get caught). It was up to the followers of Freud who revised his theories (the Neo-Freudians) to shed light on the psychoanalytic explanation of crime.

One of the first neo-Freudians to do so was August Aichorn, who took the position that it was not overdeveloped superegos but an underdeveloped superego that primarily caused crime. He believed that some criminals, raised as children without loving parents or parents at all, developed unregulated ids. Others were overindulged at the oral stage and required different treatments. In any event, Aichorn's ideas popularised the notion that delinquents needed unconditional love rather than a punitive, institutionalised setting. The ideas of maternal deprivation or love deprivation as a cause of crime are still popular.

Redl & Wineman were another group of neo-Freudians and took on the Freudian notion of Oedipus Complex. According to orthodox Freudian theory, criminals should hate their fathers more than their mothers, but Redl & Wineman found that criminals hate both their parents. In fact, they hadn't gone through a genital stage at all. Their egos were therefore undeveloped, and with nothing to mediate between the id and superego, their personalities were nothing but an endless series of raging conflicts, and this is what they called the "delinquent ego".

However these theories are based on hypothetical constructs and has little if any empirical research to support them. The social learning theory states that crime is learned behaviour. People learn criminal behaviour through the groups with which they associate. If a person associates with more groups that define criminal behaviour as acceptable than groups that define criminal behaviour as unacceptable, the person will probably engage in criminal behaviour. Just as people must learn though socialisation how to conform to their society's norms, they must also learn how to depart from those norms. In other words, deviance, like conforming behaviour, is a product of socialisation. This theory shows how a juvenile can socially learn deviant behaviour from those around him/her such as family, peers, schoolmates or anyone else that he or she may come in contact with. The parents and peers are probably the most powerful agents in socialisation.

For example, a child growing up in a home where the parents routinely engaged in criminal acts would grow up assuming that these acts may not be as wrong as society or the law has defined them. If a child is around delinquent peers, they can also learn the activities of their peers and be much more prone to engaging in criminal activity.

There is no clear theory of abnormal behaviour in relation to crime as the views on abnormal behaviour are specifically related to schools so in this instance we are simply looking at the overall psychological theories of abnormal behaviour.

Eysenck's (1987) theory of personality and crime is a famous illustration of connecting personality factors with criminal offending. Eysenck holds the view that people are hedonistic animals and that offending is a pleasurable activity. This hedonistic tendency to commit crime, however, is opposed by the development of the conscience. He suggests the conscience is a conditioned fear response and consequently the likelihood that an individual will commit crime depends on the strength of the conscience. Children who do not engage in criminal activity have developed a conscience and hence suffer guilt if they do something wrong. In contrast, children who commit criminal acts have not developed strong consciences. Eysenck contends that criminals have "poor conditionality" as linked to his 3 dimensions of personality, which are: extraversion (E), neuroticisim (N), and psychoticism (P). Extraverted individuals are active, aggressive and impulsive. Neurotic individuals are restless, emotionally volatile, and hypersensitive. Individuals high in psychoticism lack empathy and are insensitive (in a cruel manner). According to Eysenck, individuals high in (E) have low levels of cortical arousal and build up conditioned responses less well. Likewise, individuals high in (N) condition less well because their anxiety interferes with conditioning. Lastly, individuals high in (P) correlate with criminal offending. The latter groups of individuals tend to be emotionally cold, have little empathy, a high rate of hostility, and are inhumane. Generally, Eysenck's theory has shown strong support for a link between impulsivity and criminal behaviour as (E) consists of two semi-independent factors (impulsiveness and sociability).

## Eysenck's Theory of Personality and Crime

The late Hans J. Eysenck, British psychologist, is most well known for his theory on personality and crime. His theory proposed that "criminal behaviour is the result of an interaction between certain environmental conditions and features of the nervous system". This is certainly not one of the contemporary theories of crime, rather, Eysenck's emphasis is placed on the genetic predisposition toward antisocial and criminal behaviour. Followers of his theory believe that each individual offender has a unique neurophysiological makeup that when mixed with a certain environment, therefore, can't help but result to criminality (Bartol). It is important to note that Eysenck was not suggesting that criminals are born, rather that the combination of environment, neurobiological, and personality factors give rise to different types of crimes, and those different personalities were more susceptible to specific criminal activity. To further understand this theory, Eysenck explains it as follows:

It is not itself, or criminality that is innate; it is certain peculiarities of the central and autonomic nervous system that react with the environment, with upbringing, and many other environmental factors to increase the probability that a given person would act in a certain antisocial manner. Eysenck shows three main factors for temperament, being extraversion, neuroticism, and psychoticism. A large majority of crime research today focuses precisely on the first two traits. Eysenck did not actually identify psychoticism until later, when he found a need to identify behaviour that cannot be explained as extraversion and neuroticism.

Eysenck's studies showed that the typical extravert tends to lose his temper quickly, becoming aggressive and unreliable. He then believed that extraverts need a higher level of excitement and stimulation, known as "arousal theory." The need for high amounts of stimulation then lead to more likely encounters with the law. "They enjoy pranks and practical jokes and find challenge in opportunities to do the unconventional, or even to engage in antisocial behaviour" (Bartol). The physiological bases of extraversion are related to the Reticular Activating System (RAS). The RAS arouses the cerebral cortex and keeps it alert to incoming stimuli.

The base of neuroticism is frequently linked to the emotional area of the brain. This dimension reacts to how one successfully deals with stressful events. Whereas the extraversion centre of the brain is linked to the central nervous system, neuroticism relates to the autonomic nervous system. Neurotic individuals are believed to achieve an emotional level quickly and then remain at that level for a longer amount of time than non-neurotic individuals.

As mentioned earlier, Eysenck used the word psychoticism as a word to identify behaviours that are not explained by neuroticism and extraversion.

Eysenck used his research to categorize individuals who exhibited cold, cruel, unemotional, and insensitive characteristics, not the clinical definition of psychotic which means out of touch with reality.

To sum up the Eysenck's Theory of Criminality, offenders as a whole will demonstrate low levels of extraversion (cortical arousal), high levels of neuroticism (autonomic arousal), and are more tough-minded in the psychotic sense. Although there is much research that refutes this theory, researchers believe that if new data were modified, the theory as a whole may still be promising and useful.

## Psychodynamic Trait Theory

Psychodynamic (Psychoanalytical) therapy was developed by Sigmund Freud in the late 1800's and has then become a significant theory in the history of criminality (Siegel, 2005). Freud believed that every individual carries "[the]residue of the most significant emotional attachments of our childhood, which then guides our future interpersonal relationships" (Siegel). The theory is a three-part structure made up of the id, the ego, and the super ego. The id is considered the underdeveloped or primitive part of our make-up. It controls our need for food, sleep, and other basic instincts. This part is purely focused on instant gratification. The ego controls the id by setting up boundaries. The superego is in charge of judging the situation through morality.

Psychodynamic theorists believe that offenders have id-dominated personalities. In other words, they lose control of the ego and the id's need for instant gratification takes over. This causes impulse control problems and increased pleasure-seeking drives. Other problems associated with a damaged ego are immaturity, poor social skills, and excessive dependence on others. The idea is that negative experiences in an offenders childhood damages the ego, therefore, the offender is unable to cope with conventional society. (Siegel). Other psychoanalytical theorists believe that many criminals are driven by an unconscious need to be punished for previous sins (either real or imaginary). Therefore, "crime is a manifestation of feelings of oppression and people's inability to develop the proper psychological defence and rationales to keep these feelings under control (Siegel).

## Trait Theory

Trait theory is a more extreme version of Eysenck's theory. The view is that criminality is a product of abnormal biological or psychological traits. The root of trait theory can be traced back to Italian criminologist Cesare Lombroso. His research regarding trait theory is still considered historical curiosity, not scientific fact, but it is a theory none the less (Siegel, 2005). Lombroso believed that offenders were atavists. The word atavism refers to "an ancient, ancestral trait that appears in modern life." He stated, "[Criminals were] Neanderthal-like beings born, by some unexplained evolutionary glitch, into the modern world. Because offenders were considered "throwbacks to the prehistoric past," there were certain characteristics that were supposed to be identifiable. These features were considered to look more primitive and ape-like. These distinguishing characteristics were: small skulls, sloping foreheads, jutting brows, protruding ears, bad teeth, barrel chests, disproportionately long arms, and various other traits. Unfortunately, Lombroso's trait theory has been compared to the "nineteenth century pseudoscience of phrenology" (Schechter). We know today that criminals come in all shapes and sizes.

Contemporary trait theorists do not suggest that a single physical or biological attribute explains all criminality. Rather, each criminal has a unique set of characteristics that explain behaviour. The understanding is now split among many possibilities. Some may have inherited criminal tendencies, some may have neurological problems, and yet other research shows some criminals may have blood chemistry disorders which heighten antisocial activity. There is a definite link between behaviour patterns and chemical changes in the brain and nervous system.

Biocriminologists believe that criminals are genetically predetermined. They maintain that the body needs a stable amount of minerals and chemicals for normal brain functioning and growth. "Chemical and mineral imbalance leads to cognitive and learning deficits...and these factors in turn are associated with antisocial behaviour" (Schechter).

Researched studies have lined hypoglycemia to violence and abnormal levels of male sex hormones produce aggressive behaviour. Other physiological correlates of crime and antisocial behaviour are low serotonin, low autonomic arousal, and impaired prefrontal cortical functioning. Many of the genes associated with crime affect the neurotransmitter systems. "A gene that confers sensitivity to dopamine may increase sensation seeking-which is a characteristic of antisocial behaviour".

## Social Structure Theory

If biology could explain criminality, then why is the majority of crime and violence in poor, underdeveloped neighbourhoods? To ignore environmental and social aspects contributing to crime would be a mistake. People who live in the United States live in what is called a "stratified society" (Siegel) Stratification refers to, "a hierarchical arrangement...compromising three main layers: upper class, middle class, and lower class". There are three mini theories which fall under the Social Structure Theory which attempt to explain how one's environment and social circle can aid to crime.

*Social disorganization theory*: focuses on the urban conditions that effect crime rates. A disorganized area is one in which institutions of social control, such as family, commercial establishments and schools have broken down and can no longer perform their expected or stated functions. Indicators of social disorganization include high unemployment and school dropout rates, deteriorated housing, low income levels and large numbers of single parent households. Residents in these areas experience conflict and despair, and as a result, antisocial behaviour flourishes.

*Strain theory*: holds that crime is a function of the conflict between people's goals and the means they can use to obtain them. Strain theorists argue that although social and economic goals are common to people in all economic strata, the ability to obtain these goals is class-dependent...members of the lower class are unable to achieve [symbols of] success through conventional means. Lower class citizens can both accept their conditions and live socially responsible...or they can choose an alternative means of achieving success, such as theft or violence.

*Cultural deviance theory*: combines elements of both strain and social disorganization theories. Because of this view...a unique lower-class culture develops in disorganized neighbourhoods. Criminal behaviour is an expression of conformity to lower class sub-culture values and traditions, not a rebellion against traditional society.

## Social Process Theory

Social process theorists believe that criminality is a "function of individual socialization, and the interactions people have with organizations, institutions, and processes of society" (Siegel). Perhaps the most common approach to the social process theory is learning theory.

Albert Bandura, an influential psychologist of the twentieth century, was the first to experiment with this idea. His observations began with animals and showed that showed that they do not have to actually experience certain events in their environment to learn effectively. In relation to criminality, one can learn to be aggressive by observing others acting aggressively. An example being: if "A" beats up other children on the playground and steals money from the victims, his little brother "B" is observing this situation. When "A" then uses the money to buy toys, "B" witnesses his big brother getting rewarded for the violent act through purchasing fun things to play with.

In reality, it didn't matter that "A" was wrong; his behaviour resulted in a positive result.

There are two other approaches to social process theory. Social control theory is when one's behaviour is groomed through the close associations of institutions and individuals. The second is social reaction theory. If an individual is already viewed (labelled) as a criminal from an early age, then it is more likely that this person will see becoming a criminal as fulfilling a prophecy, thus beginning his criminal career (Siegel).

## Social Conflict Theory

Social conflict theorists believe a person, group, or institution has the power and ability to exercise influence and control over others (Farrington & Chertok, 1993). Conflict theorists are concerned with:
1. The role government plays in creating a crimogenic environment,
2. The relationship between personal or group power and the shaping of criminal law, 3. The prevalence of bias in justice system operations, and
4. The relationship between a capitalist, free enterprise economy and crime rates"

They define crime as "a political concept designed to protect the power and position of the upper classes at the expense of the poor (Siegel). The idea is that each society produces its own type and amount of crime. They have their own way of dealing with crime, and thus, get the amount of crime that they deserve (Siegel). In other words, to control and reduce crime, societies must change the social conditions that promote crime.

# 8

# Criminal Mind and Offender Profiling

## CRIMINAL MINDS

*Criminal Minds* is a crime drama that premiered on CBS on September 22, 2005. The show follows the adventures of a team of profilers from the FBI's Behavioral Analysis Unit (BAU) at Quantico, Virginia. It starred former *Chicago Hope* actors Mandy Patinkin as Jason Gideon, and stars Thomas Gibson as Aaron Hotchner. *Criminal Minds* differs from many criminal system procedural dramas by focusing on the criminal rather than the crime itself. On July 17, 2007, it was announced that Patinkin had been granted a request to be released from his role on the show. Joe Mantegna has signed on as his replacement. The series' third season premiered September 26, 2007.

### Themes and Plot Devices

Each episode has a voice-over in one of the early sequences, and again in the concluding one, discussing a quote by a well-known author or poet. The quote ties in to the case in the episode and by revisiting it at the end of the episode a sense of conclusion is achieved. The voice-over is most often given by Gideon, but occasionally Hotchner, Reid, Elle, and even JJ have done so, as well. These quotes, along with others, can be found here...

### Characters

#### Current Characters

*FBI BAU Supervisory Special Agent Aaron "Hotch" Hotchner* : Played by Thomas Gibson, Hotch was formerly assigned to the FBI Field Office in Seattle, Washington. Currently, he is the head of the BAU team. He and his wife are parents of a baby son. Hotch's attempts to balance his family life and his job successfully is an ongoing theme on the show. After a two-week suspension for releasing a serial killer, Hotch requested a transfer by Section Chief Erin Strauss, who was pressuring him for such a request. This pleased his wife, until Agent Morgan begged with Hotch to come help on a case in Milwaukee. Hotch decided to stay with the BAU, but when he returned to Virginia, his wife and son had left.

*FBI BAU Senior Special Agent Derek Morgan:* Played by Shemar Moore, Morgan is a confident and assertive everyman character, the son of an African American father and Causasian mother. He has a black belt in Judo, runs FBI self-defense classes, and previously served in a bomb squad unit. He and his two sisters (Sarah and Desirée) and grew up in a tough urban Chicago neighborhood. After the death of his father when he was 10, Morgan struggled somewhat, youthful fighting earned him a juvenile criminal record. He was taken under the wing of a local youth center coordinator, Carl Buford. Buford acted as a surrogate father to Morgan and helped Morgan obtain to a college football scholarship, but also sexually abused him, one episode focused on this.

*FBI BAU Special Agent Dr. Spencer Reid:* Played by Matthew Gray Gubler, Doctor Reid is a genius who graduated from high school at age 12. In his youth, his father left him and his mother, no longer able to deal with the paranoid schizophrenia of Reid's mom. Reid grew up learning nearly everything he knows from books, with his mother often reading to him. Still, Reid knew the way his mother was living wasn't healthy. When he was eighteen, he had his mother placed in a mental institution. She is still there, and Reid has stated that although he sends letters every day, he is afraid to visit her. Reid is also worried about the fact that his mother's illness can be passed on genetically; he once told Morgan that "I know what it's like to be afraid of your own mind". After being kidnapped by a serial killer with multiple personalities and tortured and drugged for days, Reid developed an addiction problem. This problem was discovered by Hotch and Gideon easily, as well as an old friend of Reid's in New Orleans. He seems to have recovered though the situation remains unconfirmed.

*FBI BAU Special Agent Emily Prentiss:* Played by Paget Brewster, Prentiss is the daughter of an ambassador. In her first appearance, she recognizes Hotch from one of his first commands: security clearance for her father. Her arrival surprises both Hotch and Gideon, as neither of them had signed off on the transfer. Prentiss insists her parents have not pulled strings for her. Ultimately, she joins the team at the end of episode 2x09 ("The Last Word") on a probationary basis. She replaces Agent Elle Greenaway. She is a graduate of Yale and has been working for the FBI for little under ten years, primarily in the Midwest. After being pressured by Section Chief Erin Strauss to assist in the end of Hotch's career, she had quit the FBI. However, Garcia managed to keep her departure on hold in the computer system just long enough for the rest of the team to convince her to stay with the BAU. However, upon her return, Section Chief Strauss remarked that neither Prentiss, Hotch, or the rest of the team would be able to climb the ladder to the top because of all this, thus shattering Prentiss's dreams.

*FBI BAU Special Agent Jennifer "JJ" Jareau:* Played by A.J. Cook, JJ often acts as the team's liaison with the media and local police agencies. She once went on a date to a Washington Redskins game with Reid, but little has come of it romantically. JJ graduated from East Allegheny High School near Pittsburgh, PA, where she was the captain of the varsity soccer team her senior year and earned an athletic scholarship to the University of Pittsburgh. When she and Reid split up the night of Reid's abduction, JJ walked into a dark barn and was attacked by several ferocious dogs, which she shot and killed. This had a great effect on her emotionally but has not come up since then.

*FBI Audio/Visual Technician Penelope Garcia:* Played by Kirsten Vangsness, Garcia is the team's Audio/Visual Technician at BAU Headquarters in Quantico, VA. When she speaks with Morgan they use pet names such as "hot stuff" or "sweet cheeks". Although they haven't pursued a romantic relationship, Garcia did show mild jealousy at the sight of Morgan dancing with two other women. She is into online games, specifically MMORPGs. She has broken down crying several times due to the fact that area of her job is to listen to and watch terrifying things in her office to analyse them for the team. However, according to Agent Hotchner, she "fills her office with figurines and colour to remind herself to smile as the horror fills her screens."

## Recurring Characters

*Haley Hotchner:* Played by Meredith Monroe, she is the wife of series regular Agent Hotchner. She and Hotch have a newborn son, Jack. Jack's name was revealed in episode 2x04 ("Psychodrama"). They are currently having marital issues and it appears that Haley has left Hotch at the end of episode 3x02 ("In Name and Blood"). Hotch mentions that he is not sure if she is coming back to Morgan at the end of episode 3x04 ("Children of the Dark").

*Agent Kramer:* Played by Gonzalo Menendez, he runs the FBI Field Office in Baltimore, Maryland, as well as the Organized Crime division in that city. As such, the two episodes which take place in Baltimore, "Natural Born Killer" (1x08) and "Honor Among Thieves" (2x20) both have him liaising with the BAU.

## Former characters

*FBI BAU Special Agent Elle Greenaway:* Played by Lola Glaudini, Elle was formerly assigned to FBI Field Office in Seattle, Washington; recently assigned to the BAU, expert in sexual offense crimes. Her father was a police officer but was killed in the line of duty. She is half Cuban and speaks Spanish. After she was shot, she returned to the BAU rather quickly, against the advice of the rest of the team.

Not long after, she staked out and shot a serial rapist in cold-blood because of his crimes against women. Her ability as a profiler was questioned by Hotchner and Gideon because of this, even though the local police force deemed it self-defense. She turned her badge and her gun over to Hotch, declaring that it wasn't "an admission of guilt."

*FBI BAU Supervisory Special Agent Jason Gideon:* Played by Mandy Patinkin, Gideon was the BAU's best profiler. He helped Morgan and Reid through their nightmares. Prior to the series, he is said to have had a "nervous breakdown" after he sent six men into a warehouse with a bomb in it. All six agents were killed, and he was heavily criticized about the event. Gideon is also skilled at chess, having continually beaten Dr. Reid. After a series of emotional cases, Gideon began to feel burnt out. The last straw was Hotch's two week suspension, for which Gideon felt responsible. He retreated to his cabin during the suspension and left a letter for Dr. Reid, whom he knew would be the one to come looking for him. When Reid arrived at the cabin, it was empty except for the letter and Gideon's badge and weapon. Gideon was last seen leaving a diner and remarking to the waitress that he didn't know where he was going or how he'd know when he got there, and subsequently driving off in his car.

## AUTOMATED FINGERPRINT IDENTIFICATION SYSTEM

Automated Fingerprint Identification System (or AFIS) is a system to automatically match one or many unknown fingerprints against a database of known prints. This is done for various reasons, not the least of which is because the person has committed a crime. With greater frequency in recent years, AFIS like systems have been used in civil identification projects. The intended purpose is to prevent multiple enrollment in an election, welfare, DMV or similar system. IAFIS, the 'I' meaning 'integrated', holds all fingerprint sets (called tenprints) collected in the US, and is managed by the FBI.

### International use

Many other nations, including Canada, the United Kingdom, Israel, Pakistan, Australia, and the International Criminal Police Organization, as well as many states and local administrative regions, have their own AFIS, which are used for a variety of purposes, including criminal identification, applicant background checks, receipt of benefits, and receipt of credentials (such as passports).

## Technology

The machine used which scans live fingerprints into AFIS is called the LiveScan Device. The process of obtaining the prints by way of laser scanning is called LiveScan. The process of obtaining prints by putting a tenprint card (prints taken using ink) is occasionally called DeadScan or CardScan. In addition to these devices, there are other devices to capture prints from crime scenes (latent prints), as well as devices (both wired and wireless) to capture one or two live fingers. The most common method of acquiring fingerprint images remains the inexpensive ink pad and paper form. Scanning forms ("fingerprint cards") in forensic AFIS complies with standards established by the FBI and NIST. To match a print, a fingerprint technician scans in the print in question, and the computer marks all minutiae points according to an algorithm. In some systems, the technician then goes over the points the computer has marked, and submits the minutiae to a one-to-many (1:N) search. Increasingly, there is no human editing of features necessary in the better commercial systems. The fingerprint image processor generally will assign a "quality measure" that indicates if the print is useful for searching.

## Types of Fingerprinting

Fingerprint matching algorithms vary greatly in terms of Type I (false positive) and Type II (false negative) errors. They also vary in terms of features such as image rotation invariance and independence from a reference point (usually, the "core", or center of the fingerprint pattern). The accuracy of the algorithm, robustness to poor image quality and the characteristics noted above are critical elements of system performance.

## Methods

Fingerprint matching has an enormous computational burden. Some larger AFIS vendors deploy custom hardware while others use highly optimized software to attain matching speed and throughput. In general, it is desirable to have, at the least, a two stage search. The first stage will generally improve access precision by use of global fingerprint characteristics such as "pattern type combinations" while the second stage is the minutiae matcher.

In any case, the search systems return results with some numerical measure of the probability of a match (a "score"). In tenprint searching using a "search threshold" parameter to increase accuracy, there should seldom be more than a single candidate unless there are multiple records from the same candidate in the database. Many systems use a broader search in order to reduce the number of missed identifications, however, and these searches can return from one to ten possible matches. Latent to tenprint searching will frequently return many - often fifty or more - candidates because of limited and poor quality input data. The validation of computer suggested candidates is usually done by a technician in forensic systems. In recent years, though, "lights-out" or "auto-confirm" algorithms produce "Identified" responses without a human operator looking at the prints, provided the matching score is high enough. "Lights-out" or "auto-confirm" is often used in civil identification systems, and is increasingly used in criminal identification systems as well.

## Main Vendors for Large Scale AFIS

There are only a few companies with a proven track record for the delivery of large AFIS. Within the industry the following are considered to be the front runners: Cogent Systems, Sagem Morpho (part of SAFRAN), Motorola,NEC and Dermalog.

## National DNA Database

A National DNA database is a database of DNA samples against which law enforcement agencies can match suspect DNA. The first national database was set up by the United Kingdom in April 1995. The second one was set up in New Zealand. France set up the FNAEG in 1998. Originally intended for sex offenders, it has since extended to include almost any criminal offender. In England and Wales, anyone arrested on suspicion of a recordable offence must submit a DNA sample to the database, which is then kept on permanent record. In Scotland, the law is different and most people are removed from the database if they are acquitted. In Sweden, only criminals who have spent more than two years in prison are recorded. In Norway and Germany, court orders are required, and are only available, respectively, for serious offenders and for those convicted of certain offences and likely to reoffend. All 50 states in the USA keep profiles of violent offenders, and a few keep profiles of suspects. Portugal has plans to introduce a DNA database of its entire population

## DNA Databases and Medicine

The database became the common meeting ground for computer scientists and molecular biologists. This is because the goal of certain projects like the genome project was to construct maps, which were built from information contributed to databases. They enabled an entirely new way of information analysis. The intrusion of computers into molecular biology shifted power into the hands of those with mathematical aptitudes and the computer savvy. However, the information gained from mapping and sequencing genetic information would very likely have ethical implications for individuals, families and society in general. There is a concern about genetic information being used in ways that affect chances of employment or chance of getting life insurance. Furthermore, secondary applications of personal, genetic information mean that citizens do not know what their genetic information will be used for.

## PSYCHOLOGICAL THEORIES OF CRIME

Psychological theories of crime begin with the view that individual differences in behaviour may make some people more predisposed to committing criminal acts. These differences may arise from personality characteristics, biological factors, or social interactions.

### Psychoanalytic Theory

According to Sigmund Freud (1856-1939), who is credited with the development of psychoanalytic theory, all humans have natural drives and urges repressed in the unconscious. Furthermore, all humans have criminal tendencies. Through the process of socialization, however, these tendencies are curbed by the development of inner controls that are learned through childhood experience. Freud hypothesized that the most common element that contributed to criminal behaviour was faulty identification by a child with her or his parents. The improperly socialized child may develop a personality disturbance that causes her or him to direct antisocial impulses inward or outward. The child who directs them outward becomes a criminal, and the child that directs them inward becomes a neurotic.

### Cognitive Development Theory

According to this approach, criminal behaviour results from the way in which people organize their thoughts about morality and the law. In 1958, Lawrence Kohlberg, a developmental psychologist, formulated a theory concerning the development of moral reasoning. He posited that there are three levels of moral reasoning, each consisting of two stages. During middle childhood, children are at the first level of moral development. At this level, the *pre-conventional* level, moral reasoning is based on obedience and avoiding punishment. The second level, the *conventional level* of moral development, is reached at the end of middle childhood.

The moral reasoning of individuals at this level is based on the expectations that their family and significant others have for them. Kohlberg found that the transition to the third level, the post-conventional level of moral development, usually occurs during early adulthood. At this level, individuals are able to go beyond social conventions. They value the laws of the social system; however, they are open to acting as agents of change to improve the existing law and order. People who do not progress through the stages may become arrested in their moral development, and consequently become delinquents.

## Learning Theory

Learning theory is based upon the principles of behavioural psychology. Behavioural psychology posits that a person's behaviour is learned and maintained by its consequences, or reward value. These consequences may be external reinforcement that occurs as a direct result of their behaviour (e.g. money, social status, and goods), vicarious reinforcement that occurs by observing the behaviour of others (e.g. observing others who are being reinforced as a result of their behaviour), and self-regulatory mechanisms (e.g. people responding to their behaviour).

According to learning theorists, deviant behaviour can be eliminated or modified by taking away the reward value of the behaviour. Hans J. Eysenck, a psychologist that related principles of behavioural psychology to biology, postulated that by way of classical conditioning, operant conditioning, and modelling people learn moral preferences. Classical conditioning refers to the learning process that occurs as a result of pairing a reliable stimulus with a response.

Eysenck believes, for example, that over time a child who is consistently punished for inappropriate behaviour will develop an unpleasant physiological and emotional response whenever they consider committing the inappropriate behaviour. The anxiety and guilt that arise from this conditioning process result in the development of a conscience. He hypothesizes, however, that there is wide variability among people in their physiological processes, which either increase or decrease their susceptibility to conditioning and adequate socialization.

## Intelligence and Crime

James Q. Wilson's and Richard J. Herrnstein's *Constitutional-Learning Theory* integrates biology and social learning in order to explain the potential causes of criminality. They argue that criminal and noncriminal behaviour have gains and losses. If the gains that result from committing the crime (e.g. money) outweigh the losses (e.g. being punished), then the person will commit the criminal act. Additionally, they maintain that *time discontinued equity* are two other variables that play an important role in criminality.

*Time discounting* refers to the immediate rewards that result from committing the crime vis-a-vis the punishment that may result from committing the crime, or the time that it would take to earn the reward by noncriminal means. Because people differ in their ability to delay gratification, some persons may be more prone to committing criminal acts than others. Moreover, judgments of *equity* may result in the commission of a criminal act. The gains associated with committing the crime may help to restore a person's feelings of being treated unjustly by society. Wilson and Herrnstein hypothesize that there are certain constitutional factors (such as intelligence and variations is physiological arousal) that determine how a person weighs the gains and losses associated with committing a criminal act.

According to Wilson and Herrnstein, physiological arousal determines the ease in which people are classically conditioned; therefore, people who are unable to associate negative feelings with committing crime will not be deterred from committing criminal acts. In addition, they argue that impulsive, poorly socialized children of low intelligence are at the greatest risk of becoming criminals. However, they have only demonstrated that low intelligence and crime occur together frequently; they have not demonstrated that low intelligence is the cause of crime.

## Serial Murder and Social Learning Theory

Hale (1993) applied the social learning theory to serial murder using case studies, and he claimed that serial murder can be learned. The social learning theory suggests that people learn new behaviour through punishment and rewards. Hale argued that serial murderers must go through some humiliating experience in the early development of their life. But the serial murderer goes through a different process because most children go through some sort of humiliation during their life.

The child who becomes a serial killer is often introduced to a humiliating experience, and cannot distinguish between a rewarding and non rewarding experience, which is part of the social learning theory. This causes the child to look at certain situations in a negative way, causing the child to become frustrated. When the individual becomes frustrated from a humiliating experience from the past, the individual then choose vulnerable outlets for their aggression. The child learns to expect humiliation or a negative situation from the past, which then causes frustration or aggression.

*Case Examples:* Ed Gein was humiliated early in his life and later turned his aggression out on others. Gein was controlled by his mother, and rejected by his father as a child, and was often abused.

Ted Bundy chose his victims based on the resemblance to a former girlfriend who had broken their marriage engagement. David Berkowitz had a sense of rejection stemmed from being adopted, and it was said he felt rejected and humiliated by the world. In this case, Berkowitz turned to fire starting the vent his frustration as a child. Later in his life, Berkowitz obtained a sexual transmitted disease which created more hatred for women, which he would later turn to kill random women. In all of these instances the serial killer was presented with some form of humiliation as a child, and learned to vent their anger through aggression.

## Applications

The applications of social learning theory have been important in the history of education policies in the United States. The zone of proximal development is used as a basis for early intervention programs such as Head Start. Social learning theory can also be seen in the TV and movie rating system that is used in the United States. The rating system is designed to all parents to know what the programs that their children are watching contain. The ratings are based on age appropriate material to help parents decide if certain content is appropriate for their child to watch. Some content may be harmful to children who do not have the cognitive ability to process certain content, however the child may model the behaviours seen on TV.

Guided participation is seen in schools across the United States and all around the world in language classes when the teacher says a phrase and asks the class to repeat the phrase. The other part to guided participation is when the student goes home and practices on their own. Guided participation is also seen with parents who are trying to teach their own children how to speak.

Portraitising is another technique that is used widely across the United States. Most academic subjects take advantage of portraitising, however mathematics is one of the best examples. As students move through their education they learn skills in mathematics that they will build on throughout their scholastic careers. A student who has never taken a basic math class and does not understand the principles of addition and subtraction will not be able to understand algebra. The process of learning math is a portraitising technique because the knowledge builds on itself over time.

## CRIMINAL PROFILING

Through the techniques used today law enforcement seeks to do more than describe the typical murderer, if in fact there ever was such a person. Investigative profiles analyse information gathered from the crime scene for what it may reveal about the type of person who committed the crime.

Law enforcement has had some outstanding investigators; however, their skills, knowledge and thought processes have rarely been captured in the professional literature. These people were truly the experts of the law enforcement field, and their skills have been so admired that many fictional characters (Sergeant Cuff, Sherlock Holmes, Hercule Poirot, Mike Hammer, and Charlie Chan) have been modeled on them. Although Lunde (1976) has stated that the murders of fiction bear no resemblance to the murders of reality, a connection between fictional detective techniques and modem criminal profiling methods may indeed exist. For example, it is attention to detail that is the hallmark of famous fictional detectives; the smallest item at a crime scene does not escape their attention. As stated by Sergeant Cuff in Wilkie Collins' The Moonstone, widely acknowledged as the first full-length detective study:

At one end of the inquiry there was a murder, and at the other end there was a spot of ink on a tablecloth that nobody could account for. In all my experience... I have never met with such a thing as a trifle yet. However, unlike detective fiction, real cases are not solved by one tiny clue but the analysis of all clues and crime patterns. Criminal profiling has been described as a collection of leads, as an educated attempt to provide specific information about a certain type of suspect, and as a biographical sketch of behavioural patterns, trends, and tendencies. Geberth (1981) has also described the profiling process as particularly useful when the criminal has demonstrated some form of psychopathology.

As used by the FBI profiles, the criminal-profile generating process is defined as a technique for identifying the major personality and behavioural characteristics of an individual based upon an analysis of the crimes he or she has committed. The profiles's skill is in recognizing the crime scene dynamics that link various criminal personality types who commit similar crimes.

The process used by an investigative profiles in developing a criminal profile is quite similar to that used by clinicians to make a diagnosis and treatment plan: data are collected and assessed, the situation reconstructed, hypotheses formulated, a profile developed and tested, and the results reported back. Investigators traditionally have learned profiling through brainstorming, intuition, and educated guesswork. Their expertise is the result of years of accumulated wisdom, extensive experience in the field, and familiarity with a large number of cases.

A profiles brings to the investigation the ability to make hypothetical formulations based on his or her previous experience. A formulation is defined here as a concept, that organizes, explains, or makes investigative sense out of information, and that influences the profile hypotheses. These formulations are based on clusters of information emerging from the crime scene data and from the investigator's experience in understanding criminal actions.

A basic premise of criminal profiling is that the way a person thinks (i.e., his or her patterns of thinking) directs the person's behaviour. Thus, when the investigative profiles analyses a crime scene and notes certain critical factors, he or she may be able to determine the motive and type of person who committed the crime.

## The Criminal-Profile-Generating Process

Investigative profiles at the FBI's Behavioural Science Unit (now part of the National Center for the Analysis of Violent Crime [NCAVC]) have been analyzing crime scenes and generating criminal profiles since the 1970s. Our description of the construction of profiles represents the off-site procedure as it is conducted at the NCAVC, as contrasted with an on-site procedure. The criminal-profile-generating process is described as having five main stages with a sixth stage or goal being the apprehension of a suspect.

## Profiling Inputs Stage

The profiling inputs stage begins the criminal-profile-generating process. Comprehensive case materials are essential for accurate profiling. In homicide cases, the required information includes a complete synopsis of the crime and a description of the crime scene, encompassing factors indigenous to that area to the time of the incident such as weather conditions and the political and social environment. Complete background information on the victim is also vital in homicide profiles. The data should cover domestic setting, employment, reputation, habits, fears, physical condition, personality, criminal history, family relationships, hobbies, and social conduct.

Forensic information pertaining to the crime is also critical to the profiling process, including an autopsy report with toxicology/serology results, autopsy photographs, and photographs of the cleansed wounds. The report should also contain the medical examiner's findings and impressions regarding estimated time and cause of death, type of weapon, and suspected sequence of delivery of wounds.

In addition to autopsy photographs, aerial photographs (if available and appropriate) and 8 x 10 colour pictures of the crime scene are needed. Also useful are crime scene sketches showing distances, directions, and scale, as well as maps of the area (which may cross law enforcement jurisdiction boundaries).

In some cases, murder may be an ancillary action and not itself the primary intent of the offender. The killer's primary intent could be:
(1) criminal enterprise.
(2) emotional, selfish, or cause-specific, or
(3) sexual. The killer may be acting on his own or as part of a group.

When the primary intent is criminal enterprise, the killer may be involved in the business of crime as his livelihood. Sometimes murder becomes part of this business even though there is no personal malice toward the victim.

The primary motive is money. In the 1950s, a young man placed a bomb in his mother's suitcase that was loaded aboard a commercial aircraft. The aircraft exploded, killing 44 people. The young man's motive had been to collect money from the travel insurance he had taken out on his mother prior to the flight. Criminal enterprise killings involving a group include contract murders, gang murders, competition murders, and political murders.

When the primary intent involves emotional, selfish, or cause-specific reasons, the murderer may kill in self-defense or compassion (mercy killings where life support systems are disconnected). Family disputes or violence may lie behind infanticide, matricide, patricide, and spouse and sibling killings. Paranoid reactions may also result in murder as in the previously described Whitman case. The mentally disordered murderer may commit a symbolic crime or have a psychotic outburst. Assassinations, such as those committed by Sirhan Sirhan and Mark Chapman, also fall into the emotional intent category. Murders in this category involving groups are committed for a variety of reasons: religious (Jim Jones and the Jonestown, Guyana, case), cult (Charles Manson), and fanatical organizations such as the Ku Klux Klan and the Black Panther Party of the 1970s.

Finally, the murderer may have sexual motives for killing. Individuals may kill as a result of or to engage in sexual activity, dismemberment, mutilation, evisceration, or other activities that have sexual meaning only for the offender. Occasionally, two or more murderers commit these homicides together as in the 1984-1 985 case in Calaveras County, California, where Leonard Lake and Charles Ng are suspected of as many as 25 sex-torture slayings.

## Victim Risk

The concept of the victim's risk is involved at several stages of the profiling process and provides information about the suspect in terms of how he or she operates. Risk is determined using such factors as age, occupation, lifestyle, physical stature, resistance ability, and location of the victim, and is classified as high, moderate, or low. Killers seek high-risk victims at locations where people may be vulnerable, such as bus depots or isolated areas. Low-risk types include those whose occupations and daily lifestyles do not lead them to being targeted as victims. The information on victim risk helps to generate an image of the type of perpetrator being sought.

## Offender Risk

Data on victim risk integrates with information on offender risk, or the risk the offender was taking to commit the crime. For example, abducting a victim at noon from a busy street is high risk. Thus, a low-risk victim snatched

# Criminal Mind and Offender Profiling

under high-risk circumstances generates ideas about the offender, such as personal stresses he is operating under, his beliefs that he will not be apprehended, or the excitement he needs in the commission of the crime, or his emotional maturity.

## Escalation

Information about escalation is derived from an analysis of facts and patterns from the prior decision process models. Investigative profiles are able to deduce the sequence of acts committed during the crime.

From this deduction, they may be able to make determinations about the potential of the criminal not only to escalate his crimes (e.g., from peeping to fondling to assault to rape to murder), but to repeat his crimes in serial fashion. One case example is David Berkowitz, the Son of Sam killer, who started his criminal acts with the nonfatal stabbing of a teenage girl and who escalated to the subsequent .44-caliber killings.

## Time Factors

There are several time factors that need to be considered in generating a criminal profile. These factors include the length of time required: (1) to kill the victim, (2) to commit additional acts with the body, and (3) to dispose of the body.

The time of day or night that the crime was committed is also important, as it may provide information on the lifestyle and occupation of the suspect (and also relates to the offender risk factor).

For example, the longer an offender stays with his victim, the more likely it is he will be apprehended at the crime scene. In the case of the New York murder of Kitty Genovese, the killer carried on his murderous assault to the point where many people heard or witnessed the crime, leading to his eventual prosecution. A killer who intends to spend time with his victim therefore must select a location to preclude observation, or one with which he is familiar.

## Location Factors

Information about location-where the victim was first approached, where the crime occurred, and if the crime and death scenes differ-provide yet additional data about the offender. For example, such information provides details about whether the murderer used a vehicle to transport the victim from the death scene or if the victim died at her point of abduction.

## PSYCHOLOGICAL PROFILING

In 1957, the identification of George Metesky the arsonist in New York City's Mad Bomber case (which spanned 16 years) was aided by psychiatrist-criminologist James A. Brussel's staccato-style profile: "Look for a heavy man. Middle-aged Foreign born. Roman Catholic. Single. Lives with a brother or sister. When you find him chances are he'll be wearing a double-breasted suit. Buttoned."

Indeed the portrait was extraordinary in that the only variation was that Metesky lived with two single sisters. Brussel in a discussion about the psychiatrist acting as Sherlock Holmes explains that a psychiatrist usually studies a person and makes some reasonable predictions about how that person may react to a specific situation and about what he or she may do in the future. What is done in profiling according to Brussel is to reverse this process. Instead, by studying an individual's deeds one deduces what kind of a person the individual might be.

The idea of constructing a verbal picture of a murderer using psychological terms is not new. In 1960 Palmer published results of a three-year study of 51 murderers who were serving sentences in New England. Palmer's "typical murderer" was 33 years old when he committed murder. Using a gun, this typical killer murdered a male stranger during an argument. He came from a low social class and achieved little in terms of education or occupation. He had a well meaning but maladjusted mother and he experienced physical abuse and psychological frustrations during his childhood.

Similarly, Rizzo (1982) studied 31 accused murderers during the course of routine referrals for psychiatric examination at a court clinic. His profile of the average murderer listed the offender as a 26-year-old male who most likely knew his victim with monetary gain the most probable motivation for the crime.

## CRIMINAL PROFILE

Based on the information derived during the previous stages, a criminal profile of the murderer was generated. First, a physical description of the suspect stated that he would be a white male, between 25 and 35, or the same general age as the victim, and of average appearance. The murderer would not look out of context in the area. He would be of average intelligence and would be a high-school or college dropout. He would not have a military history and may be unemployed. His occupation would be blue-collar or skilled. Alcohol or drugs did not assume a major role, as the crime occurred in the early morning.

The suspect would have difficulty maintaining any kind of personal relationships with women. If he dated, he would date women younger than himself, as he would have to be able to dominate and control in the relationships. He would be sexually inexperienced, sexually inadequate, and never married. He would have a pornography collection. The subject would have sadistic

# Criminal Mind and Offender Profiling

tendencies; the umbrella and the masturbation act are clearly acts of sexual substitution. The sexual acts showed controlled aggression, but rage or hatred of women was obviously present. The murderer was not reacting to rejection from women as much as to morbid curiosity.

In addressing the habits of the murderer, the profile revealed there would be a reason for the killer to be at the crime scene at 6:30 in the morning. He could be employed in the apartment complex, be in the complex on business, or reside in the complex. Although the offender might have preferred his victim conscious, he had to render her unconscious because he did not want to get caught. He did not want the woman screaming for help.

The murderer's infliction of sexual, sadistic acts on an inanimate body suggests he was disorganized. He probably would be a very confused person, possibly with previous mental problems. If he had carried out such acts on a living victim, he would have a different type of personality. The fact that he inflicted acts on a dead or unconscious person indicated his inability to function with a live or conscious person.

The crime scene reflected that the killer felt justified in his actions and that he felt no remorse. He was not subtle. He left the victim in a provocative, humiliating position, exactly the way he wanted her to be found. He challenged the police in his message written on the victim; the messages also indicated the subject might well kill again.

## Investigation

The crime received intense coverage by the local media because it was such an extraordinary homicide. The local police responded to a radio call of a homicide. They in turn notified the detective bureau, which notified the forensic crime scene unit, medical examiner's office, and the county district attorney's office. A task force was immediately assembled of approximately 26 detectives and supervisors. An intensive investigation resulted, which included speaking to, and interviewing, over 2,000 people. Records checks of known sex offenders in the area proved fruitless. Hand writing samples were taken of possible suspects to compare with the writing on the body. Mental hospitals in the area were checked for people who might fit the profile of this type killer.

The FBI's Behavioural Science Unit was contacted to compile a profile. In the profile, the investigation recommendation included that the offender knew that the police sooner or later would contact him because he either worked or lived in the building. The killer would somehow inject himself into the investigation, and although he might appear cooperative to the extreme, he would really be seeking information. In addition, he might try to contact the victim's family.

## Apprehension

The outcome of the investigation was apprehension of a suspect 13 months following the discovery of the victim's body. After receiving the criminal profile, police reviewed their files of 22 suspects they had interviewed. One man stood out.

This suspect's father lived down the hall in the same apartment building as the victim. Police originally had interviewed his father, who told them his son was a patient at the local psychiatric hospital. Police learned later that the son had been absent without permission from the hospital the day and evening prior to the murder.

They also learned he was an unemployed actor who lived alone; his mother had died of a stroke when he was 19 years old (11 years previous). He had academic problems of repeating a grade and dropped out of school. He was a white, 30-year-old, never-married male who was an only child.

His father was a blue-collar worker who also was an ex-prize fighter. The suspect reportedly had his arm in a cast at the time of the crime.

A search of his room revealed a pornography collection. He had never been in the military, had no girlfriends, and was described as being insecure with women. The man suffered from depression and was receiving psychiatric treatment and hospitalization. He had a history of repeated suicidal attempts (hanging/asphyxiation) both before and after the offense.

The suspect was tried, found guilty, and is serving a sentence from 25 years to life for this mutilation murder. He denies committing the murder and states he did not know the victim.

Police proved that security was lax at the psychiatric hospital in which the suspect was confined and that he could literally come and go as he pleased. However, the most conclusive evidence against him at his trial were his teeth impressions.

Three separate forensic dentists, prominent in their field, conducted independent tests and all agreed that the suspect's teeth impressions matched the bite marks found on the victim's body.

The profiles studies ail this background and evidence information, as well as all initial police reports. The data and photographs can reveal such significant elements as the level of risk of the victim, the degree of control exhibited by the offender, the offender's emotional state, and his criminal sophistication.

Information the profiles does not want included in the case materials is that dealing with possible suspects. Such information may subconsciously prejudice the profiles and cause him or her to prepare a profile matching the suspect.

## OFFENDER TYPOLOGIES

Criminologists have developed typologies of both adult and juvenile offenders. Some schemes rest on psychological criteria, whereas others use patterns of behaviour common in correctional institutions to establish criminal types; sociological approaches emphasize individual criminal activities, personal attitudes, self-concepts, group relations, and similar variables. Examples of these approaches follow.

Psychological typologies. Psychiatrist Richard Jenkins and sociologist Lester Hewitt put forth an influential psychological typology of juvenile offenders many years ago. They examined youths in a child guidance clinic and concluded that delinquents fall into two groups: pseudosocial boys and unsocialized aggressive youths. The former were psychologically normal youngsters who were responding to antisocial conditions in their local communities, while the latter were asocial, violent juveniles who had suffered severe parental rejection. more recent and well-known psychological typology of offenders is Marguerite Warren's Interpersonal Maturity Levels (I-Levels) description of delinquents. Warren hypothesized that children become well-adjusted social beings by passing successfully through seven stages, from infantile dependence to adult maturity and interpersonal competence. Some individuals fail to attain the highest levels of personal development: their development stops at an intermediate stage, and they consequently behave in relatively immature ways. According to Warren, juvenile delinquents are usually found in three of the lowest levels of maturity. For example, some are easily led by peers into misbehavior, whereas others have great difficulty conforming to reasonable demands of authority figures.

The I-Levels model is faulty in some respects (Gibbons, 1970). For example, Warren did not adequately compare nondelinquents with offenders in terms of interpersonal competence. Moreover, the validity of the scheme is suspect. Investigators have tried to assign delinquents to Warren's typology through the use of personality tests or inventories, but have generally failed to obtain results consistent with I-Levels diagnoses based on clinical judgment.

Another psychological typology rests on felons' Minnesota Multiphasic Personality Inventory scores (Megargee et al.). This typology can be useful in correctional programs, but many nonoffenders would probably exhibit psychological profiles similar to those that have been found among prisoners.

Finally, a number of psychiatrists have drawn upon their clinical experiences with offenders to create less formal typologies of lawbreakers. These descriptive typologies of murderers, sex offenders, and other kinds of criminals usually emphasize forms of psychological maladjustment that are said to differentiate criminal types.

Inmate social role typologies. Existential types noted among inmates of both juvenile correctional institutions and adult facilities have sometimes served as the basis of offender typologies. Proponents of this approach argue that these types parallel behaviour patterns among lawbreakers at large.

In this tradition, Clarence Schrag reported that male prisoners are labeled in inmate argot as "square Johns," "right guys," "outlaws," or "politicians," according to the nature of their relationships with fellow prisoners (pp. 346–356). Loyalty to other inmates is the decisive variable: thus, the "right guy" is faithful to his peers and hostile toward guards and other authority figures, whereas the "square John" is an alien in the convict social system. Schrag also contended that the criminal records and other characteristics of prisoners vary predictably with their position in the inmate role system.

In fact, however, there is only a relatively loose fit between this inmate typology and the real world. Peter Garabedian conducted research in the same prison from which Schrag's conclusions were derived. Garabedian assigned inmates to role types on the basis of their responses to an attitude questionnaire containing statements designed to reflect social roles. Significantly, he found that about one-third of the sample was unclassifiable through this procedure and that the correlations between offender characteristics and social roles were relatively weak. Robert Leger studied inmate social categories as well. Using Garabedian's attitude questionnaire to identify inmate role types, Leger asked prisoners to indicate their own type. Additionally, he queried guards concerning prisoner roles, and included social background information on inmates as another measure of role. These techniques did not consistently assign particular individuals to a single type, nor did they agree on the total number of prisoners within each role.

Sociological typologies. The state and federal criminal codes represent a typology into which lawbreakers might be placed. For example, offenders could be classified on the basis of the crime with which they are currently charged. However, criminal codes do not meet the parsimony criterion; in addition, offenders violate different laws over time, which challenges the requirement of mutual exclusivity.

Sociological criminologists have attempted to overcome such problems with legal codes by developing offender typologies that assign persons who engage in similar collections of offenses to particular criminal behaviour systems, role careers, or criminal behaviour patterns. Typological efforts have also sought to discover offender groupings whose members share social background factors and causal experiences. In short, criminologists have tried to identify sociologically meaningful types.

One of these sociological typologies was constructed by Marshall Clinard and Richard Quinney, who used both offence-and person-centred criteria to define nine criminal behaviour systems, including those involving violent personal criminal behaviour, public-order criminal behaviour, and occasional property criminal behaviour. They also discussed the criminal careers of offenders who fell into these groupings, implying that individuals in fact specialize to some degree.

Daniel Glaser has also offered a typological description of offenders. He identified ten offender patterns delimited by "offence descriptive variables" and "career commitment variables." Among his types, the "adolescent recapitulator" engages in an assortment of offenses; his criminality reflects a failure to assume a stable adult role. Glaser's "vocational predators" pursue crime as a livelihood. In this typology, drug addicts constitute another category, that of "addicted performers".

Glaser's scheme exhibits a number of flaws. First, it omits certain offenders, such as political criminals involved in political protest, and perpetrators of relatively petty offenses that are sometimes termed folk crimes—for example, traffic violations or fish and game law violations. Second, the adolescent recapitulator and certain other types are based on causal processes separating these offenders, rather than on their offence patterns and criminal careers. Third, although his descriptions of the types provide considerable detail consistent with much research on lawbreakers, Glaser's typology does not spell out the identifying characteristics of the offender categories with sufficient precision; one would be hard put to assign a population of actual offenders to these types.

Don Gibbons developed detailed and comprehensive typologies for both juvenile delinquents and adult offenders. These typologies establish offender categories on the basis of the offenses in which the person is currently involved, his or her criminal career or prior criminal record, and the self-concept and role-related attitudes of the lawbreaker. For instance, one adult offender type, the "naïve check forger," writes bad checks on his or her own bank account, shows little criminal skill, views himself or herself as a noncriminal, and expresses such views as "you can't kill anyone with a fountain pen." Some of Gibbons's types rest on research findings reported by criminologists: Edwin Lemert's study of naïve check forgers provided some of the information on which Gibbons based this type description. Other types are grounded in criminological theory. In all, Gibbons established nine delinquent and fifteen adult offender categories.

What portion of the offender population falls within these types? Although no research has directly investigated the juvenile typology, the available data seem inconsistent with Gibbons's type descriptions (Gibbons and Krohn). The most striking evidence on delinquent behaviour emphasizes its transitory or episodic character. Self-reported or so-called hidden delinquents most commonly indicate that they have been involved in infrequent and petty acts of lawbreaking, while juvenile offenders who turn up in the juvenile justice system often engage in numerous illegal acts rather than a few forms of lawbreaking. Researchers have been able to identify some of the latter as "serious and/or violent offenders" or as "chronic delinquents," but have not been able to make finer distinctions among them (Loeber, Farrington, and Waschbusch). In short, the delinquency typology posits more patterning of behaviour among juvenile offenders than actually exists.

Gibbons's adult offender typology has been directly examined, and, as will be presently shown, other studies of offenders that bear on the question of empirical congruence are also relevant to this analysis.

Clayton Hartjen and Gibbons asked county probation officers to sort probationers into Gibbons's typology categories. The raters used abridged typological profiles of offenders; moreover, groups of three officers acted as independent judges who read the probationers' case files in order to determine to which types, if any, these persons belonged. Slightly less than half of the probationers were assigned to a type, although Hartjen subsequently placed most of the unassigned cases into seven ad hoc groups.

James McKenna also studied Gibbons's adult offender typology. Having examined their arrest records, he classified inmates in a state correctional institution into twelve offender types. McKenna then sought to determine whether the combinations of characteristics said to differentiate types actually existed among offenders. In only one of the twelve types did the hypothesized pattern of behaviour and social-psychological characteristics emerge; that is, the vast majority of prisoners assigned to a particular type did not consistently exhibit similar attitudes and other differentiating features. These findings indicate that many real-life offenders cannot be assigned with precision in existing criminal typologies.

Other research studies as well have raised doubts about the accuracy of offender typologies. Joan Petersilia, Peter Greenwood, and Marvin Lavin studied forty-nine individuals serving prison terms for armed robbery. Their average age was thirty-nine, and thus most had been criminally active for some years. When these offenders were asked the number of times they had committed any of nine specific crimes since becoming involved in lawbreaking, they confessed to more than 10,000 offenses, or an average of 214 per person. Even more striking was the diversity of offenses: 3,620 drug sales, 2,331 burglaries, 1,492 automobile thefts, 995 forgeries, 993 major thefts, and 855 robberies. This finding undermines the assumption of offence specialization among lawbreakers and seriously challenges most offender typologies.

Another study by Mark Peterson, Harriet Braker, and Suzanne Polich involved over six hundred prison inmates who were asked to indicate the number and kinds of crimes they had committed during the three years prior to their present incarceration.

For each type of crime, most of those who reported doing it said they did so infrequently, but a few of them confessed to engaging in the crime repeatedly. The researchers concluded that there are two major kinds of offenders: occasional criminals and broadly active ones. Few prisoners claimed to have committed a single crime at a high rate; thus there were few crime specialists in prison.

In still another investigation, Jan and Marcia Chaiken studied 2,200 jail and prison inmates in three states who were given a lengthy questionnaire that included self-report crime items. They were then asked to report the number of times they had committed robberies, assaults, burglaries, forgeries, frauds and thefts, and drug deals. About 13 percent of them said they had committed none of the offenses in the previous two years, while the rest reported involvement in one or more of the offenses, but most commonly, at relatively low rates. Also, while a sizable minority of the prisoners admitted having frequently committed one or more crimes prior to incarceration, most of them reported having been involved in crime-switching, rather than crime specialization.

Female offenders. Until relatively recently, criminologists paid relatively little attention to types of juvenile or adult female offenders. An implicit assumption has been that most female delinquents come to the attention of the police or the juvenile court, either because of minor misbehavior or sexual promiscuity. Similarly, a common presumption has been that many adult women offenders are petty thieves such as shoplifters while others have been involved in domestic violence, often directed at their common law or legal spouses.

However, in recent years, investigators have begun to examine female lawbreaking in more detail. In particular, considerable attention has focused on the different routes or pathways through which women enter into lawbreaking. One case in point is Eleanor Miller's research on women who have been involved in "street hustling," that is, a pattern of property crimes, drug use, and prostitution. She reported that there are four pathways that lead to street hustling.

Mary Gilfus has provided parallel findings regarding women who have been involved in street crime. Similarly, Kathleen Daly has sorted female felons into several types, based on the kinds of offenses they have committed and their relationships' with spouses, boyfriends, and other associates. Then, too, investigators have directed attention at females who have been involved in homicides. Taken together, these studies indicate that there are a number of patterns or types that are involved in female criminality.

# 9
# Crime and Forensic Science

Before talking about what *forensic* and *criminal psychology* is, we must define criminal behaviour first. Criminal behaviour suggests a large number and variety of acts. Andrew and Bonta (1998) suggest four broad definitions of criminal behaviour and the acts and behaviours that fit within these domains. These four areas are legal *criminal* behaviour or actions that are prohibited my the state and punishable under the law, ethical criminal behaviour which refers to actions that violate the norms of religion and ethicality and are believed to be punishable by a supreme spiritual being, social criminal behaviour which refers to actions that violate the norms of custom and tradition and are punishable by a community and finally psychological criminal behaviour that refers to actions that may be rewarding to the actor but inflict pain or loss on others-it is criminal behaviour that is anti-social behaviour.

A good working definition can be seen as: "*antisocial* acts that place the actor at risk of becoming the focus of the attention of the criminal and juvenile professionals (Andrews and Bonta, 1998). It is difficult to define criminal behaviour as ideas of what is considered imethical, unconventional, illegal or antisocial is not stable over time or place. For example, not wearing seatbelt, homosexual activity or spanking a child are all items that have been considered both illegal or legal at one point in time.

*Delinquency* must be distinguished from criminality. Delinquency is defined as behaviour that is illegal, imethical or deviant with respect to societal values. Criminality on the other hand is defined as a breaking of existing laws, there is little or no confusion as to what constitutes illegal and legal behaviours. When measuring criminal behaviour we are trying in a way to predict future criminal behaviours. We may measure criminal behaviour by arrests and charges, however not everyone charged is found guilty. We can also measure the amounts of convictions and incarcerations. We can also measure the amount of self-reported offences and some believe that this may be a more accurate way to measure criminal behaviour. This is debatable as there may be reasons that the individuals participating in the anonymous self-report surveys may distrust that their responses are anonymous. As well, individuals may over or underestimate their crimes for personal reasons. Therefore, when we study criminal behaviour we typically study what is known about persons who have been defined as criminals through the criminal justice system.

Estimates of actual *crime rates* are usually obtained from official sources, yet different sources may yield different estimates. Crime reports generally categorize crimes by type of crime and offender characteristics such as age, gender, race and location.

By estimating crime rates we want to predict *recidivism*. Recidivism is the future criminal offences following a previous offence and is defined through new arrests, new charges and new convictions. In general the best single predictor of recidivism is number and type of previous criminal offences, and these rates will vary due to age, gender and type of crime. Items such as prior youth or adult offences, present offences, charges or and arrests under the age of 16. As well, offence history is important-such as the use of a weapon, an assault on authority, sexual assault and impaired driving-in predicting future offences.

A risk factor for criminality is anything in a persons psychology, growthal or family history that may increase the likelihood that they will become involved in some point in criminal activities. A protective factor is anything in a persons biology, psychology, growthal or family history that will decrease the likelihood that they will become involved in criminal activity.

Risk factors generally include: lower class origin, family of origin, poor personal temperament, lower aptitude, early behavioural histories, poor parenting, school based factors, poor educational/vocational/socioeconomic achievement, poor interpersonal relationships, antisocial associates which support crime, antisocial attitudes/values/beliefs and feelings and psychopathology.

Delinquency also has a relationship to attachment to peers. Brownfield and Thompson (1991) found that boys who had committed delinquent acts were more likely to have friends who had done so as well. This increased when the friend was considered a "best friend." As well, risk factors can be divided into the various childhood time periods that they may occur and increase later criminal activity. Patterson, De Baryshe and Ramsey (1989) proposed the Patterson model. Risk factors that may be experienced in early childhood include poor parental discipline and monitoring, poor family environment, coercive parent-child relationships, instability in the family and early childhood conduct problems. Middle childhood risk factors include rejection by normal or conventional peers, and academic failure and underachievement. Late childhood and early adolescence risk factors include a commitment to a deviant peer group.

The "cycle of violence hypothesis" predicts that abused children will become abusers and victims of violence will become violent offenders. Abused and neglected children are more likely to exhibit delinquent characteristics as well as criminal and violent behaviour as adults.

Some possible factors about the individual that may lead to crime can be speculated. Things such as chromosomal patterns, physiological under arousal, hyperactivity/ADHD, impulsivity, learning disabilities, fetal alcohol syndrome and other neurological damage, low IQ, hormonal imbalances, antisocial personality disorder,*psychopathology* and biochemical imbalances may contribute to risk of criminal behaviour.

## INVESTIGATION OF CRIME

Section 25 of the Indian Evidence Act states that confessions made to a police officer while in custody are inadmissible. The provision is intended to prevent police from torturing suspects to gain confessions. But Section 27 of the Indian Evidence Act allows evidence obtained through confessions into court through a back door. For example, if a suspected thief is tortured into telling the police the whereabouts of stolen property or other facts not amounting to a confession, the police can then seize the stolen property or investigate the other facts and use this evidence against the accused. The justification of torture to obtain admissible evidence under Section 27 of the Indian Evidence Act is well known and acknowledged by senior police officers, as Inspector General I. Ravi Arumugam observed in his 1994 paper:

> *In [the] Indian Evidence Act, section 25 says that, "No confession made to a police officer shall be proved as against a person accused of any offence." This clearly shows... how a policeman is regarded in our country. Until this suspicious concept is removed from the law book, the police may harp on getting some evidence for the "hard-nut to crack criminals" under section 27 IEA, they are forced to use violence to extract information in a case.*

However, instead of arguing that torture should be eliminated from investigations or that Section 27 should be amended or removed from the Indian Evidence Act because it leads to torture, Inspector-General Arumugam argues that Section 25 should be eliminated, thereby removing a suspect's only legal protection from torture when evidence is being gathered by police.

The police routinely torture children in order to obtain both evidence and confessions, as the case of Shantanu illustrates.

Shantanu, along with a friend, committed a robbery at a wedding in Bombay on December 31, 1992. Eight days later, on January 8, 1993, two plainclothes police officers, Sheikh Sahab and Arun Hawaldar, came to Shantanu's house at about 10:00 a.m. and took him to the D.N. Nagar Police Station in Bombay. The following is the account Shantanu gave of his case:

> At the station Shantanu was taken to a "separate enquiry room" where he was asked to put his hands on a table, palms up, and was then beaten with a police baton on the hands. After about an hour, the person who filed the complaint came to the station and identified Shantanu as the robber. Once identified, Shantanu was taken back to the enquiry room where the police hung him from the ceiling and proceeded to beat him for about forty-five minutes with police batons on the shoulders, back, and thighs. Following this, he was forced to lie on a block of ice while his legs were held in place by police. The police hit him whenever he tried to move. Then he was taken outside and made to lie in the sun for two hours while police asked him where the stolen property was kept. When he said that he did not know where the property was, the police beat him.

Two days later, while Shantanu was still in custody, the police brought his parents to the station and threatened to beat his parents if he did not confess to the robbery. He confessed, the stolen property was recovered, and the police released his parents. The police then beat him and told him to confess to any crimes he had committed in the past. He was kept in the police station for ten more days. During that time, Sheikh Sahab would take Shantanu to the beach and ask him to escape so that he could shoot him. Shantanu claimed that Sahab tortured him because Shantanu had once fought with Sahab's brother.

On January 18, 1993, Shantanu was produced before the 10th Metropolitan Magistrates Court at Andheri in Bombay. The police falsely recorded his age as nineteen, even though he was fifteen. The police brought him before the court with his face covered and he could not hear the court proceedings. Shantanu reported that he had been tortured so badly that he could not stand, and two policemen had to hold him up during the proceedings. Despite this, he was remanded to police custody for eight days, during which the police beat him for five days. Then he was sent to the Bombay Central Prison.

The inspector-general of the prison was not convinced that Shantanu was an adult and asked the magistrate to provide age verification. Upon seeing his birth certificate, the inspector-general sent him to an observation home where he was subjected to frequent beatings by the house master. One day before his release, Shantanu struck the house master, which led to his remand being extended by one month.

Upon his release, when Shantanu was presented before the juvenile court, the magistrate asked if he had been mistreated by police. Shantanu stated that he had and as a result Sheikh Sahab was suspended and Shantanu's case was dismissed. Shantanu spent the next twenty days in the hospital as a result of the torture he received in police custody and in the observation home.

Police also torture possible witnesses, accomplices, or people who were near a crime scene into revealing the identity and location of the perpetrator. Although it is illegal under Section 160 of the Code of Criminal Procedure to detain males under the age of fifteen and females of any age who are not suspects, this tactic is extremely common. On January 11, 1995, between 2:00 and 3:00 a.m., six boys, aged twelve to fourteen, who had been sleeping at the Triplicane Railway Platform in Madras were detained by two plainclothes police officers. These police officers are known as "crime police"; they are plainclothes officers in charge of investigating theft and robbery.

The boys told Human Rights Watch that the police officers woke the boys up and took them in a jeep to the D-1 Police Station in Triplicane, Madras. When they arrived at the police station, the boys were placed in the station's holding cell or "crime cell." There were three older boys (possibly seventeen or older) in the cell with them. A total of nine people were in the twelve-by-fifteen-foot cell, which had no toilet. Rajiv told us that the police threatened to put him in the "minor jail" (an observation home) if he did not confess. Vikram told us that when he woke up and saw the police with lathis, he feared he was going to be beaten. After about three hours, at approximately 6:00-6:30 a.m., three plainclothes officers entered the cell and brought a table with them. They forced all nine detainees to place their hands, palms down, on the table and proceeded to beat the boys' hands with lathis. Then they made the boys turn their hands over and proceeded to beat their palms.

The boys told Human Rights Watch that the beatings continued for approximately thirty to forty-five minutes. The police then stopped and told the boys that some auto parts, a gas cooking stove, and 10,000 rupees had been reported stolen from a lawyer's house which was near the Triplicane railway platform. They told the boys to confess or to tell them who committed the theft. According to Vikram, the police said, "Only if we beat you will you tell us the truth. If you do not tell the truth, we will beat you more."

Then the police made four of the boys sit down "ladam" style, a position in which a boy sits on the ground with his hands behind his back and his legs straight out in front. The police then beat the four boys with lathis on their legs, feet, and soles of their feet. The other five boys, who were made to sit in a corner of the cell and watch, were periodically beaten with fists and lathis.

Vikram told Human Rights Watch:
*They made me and some others sit down ladam. Then they said: "You better tell the truth in five minutes or we will beat you more." Then they just started*

to beat my feet and legs with lathis. I kept telling them that I was in a Red
Cross program learning horticulture training. I kept telling them that I didn't
know anything, but they kept on yelling at us and beating us.

Suresh told us:

I told them that I was a ragpicker and didn't do such things. I told them
to call the NGO and they [the police] would see that I was telling the truth.
But they just kept beating us. Even when we said we didn't know anything,
they just beat us.

The beatings continued until approximately 8:30 a.m., when mothers of the boys began to arrive at the police station. Neighbours had apparently told the boys' mothers that they were at the police station. The mothers came individually and were allowed to feed the boys. Vikram asked his mother to contact the NGO he was affiliated with in order to help secure his release.

At about 10:30 a.m., the police told the mothers to leave. The boys told Human Rights Watch that police then proceeded to beat the children with their fists and lathis. They continued to tell them to confess and threatened them with more torture. Rajiv told Human Rights Watch:

When my mom left, they came back in [the cell] and told us to "Tell the truth."
They yelled at us and called us filthy things and beat us. One policeman with
a bald head and a short lathi screamed and beat us very hard.

Vikram added:

They kept beating us and told us to confess. They called us "fucker" and other
filthy things. Another policeman told them, "What is this, you shouldn't hit
such small boys." This made them stop, but when he left, they beat us even
harder.

According to the boys, this treatment continued until approximately 12:30 p.m. The police then stopped and left the boys in the cell. No case had been registered. At approximately 3:30 p.m., an NGO representative arrived and demanded that the boys be released. He had been informed of the boys' whereabout by some of the boys' mothers.

The boys were released at approximately 6:00 p.m. All the boys were fingerprinted, their names and addresses were recorded, and the reason they were taken into custody was recorded. No charges were filed. Vikram told us that it was very difficult to walk because of the pain in his legs, so he and the other boys had to hold each other and walk slowly, but he was "glad to leave the station."

The NGO representative who saw these boys at the police station told Human Rights Watch:

Whatever the police want from the children, you can always guarantee they
will beat them. For our boys, they weren't beaten that badly.

The children said that they were kept in police custody for approximately fifteen to sixteen hours in total, of which they reported that they spent approximately four-and-a-half hours being beaten.

Human Rights Watch was told by several NGO representatives and lawyers that sometimes the police are aware that their methods are illegal and as a result torture children in ways that are less conspicuous. An NGO representative in Bangalore explained:

*The types of injuries seen are those associated with severe beatings. The police are careful not to go too far and attract public attention. They won't beat the head or face because then the magistrate might be suspicious, or the public. Some children are chained for days at a time.*

Sanjiv's experience is indicative of this. He was seventeen at the time of this incident. We interviewed a lawyer working on his case on December 12, 1995, who provided us with Sanjiv's statement:

*Sanjiv dropped out of school after the fourth grade and was living in Matunga, Bombay with his parents. One of his friends, Yusuf, was involved in thefts and would frequently give stolen items to Sanjiv for safe-keeping until Yusuf claimed them. Sanjiv's parents were not happy with his activities, so Sanjiv slept on the streets, outside his house.*

On March 21, 1991, four police officers from the Matunga Police Station had come to his neighbourhood inquiring about a theft. Several boys in the area told the police that Sanjiv might be involved in the crime and told them where Sanjiv lived. At approximately 2:30 a.m. on March 22, 1991, Sanjiv and two of his acquaintances, Kanhaiye and Vijay, nineteen and seventeen respectively, were taken to the Matunga Police Station.

The police made Sanjiv lie on his stomach on a bench and tied him down. Then they beat him on his feet and back while the police told him to give them the whereabouts of Yusuf, who was wanted in connection with a theft.

Sanjiv refused to reveal Yusuf's whereabouts, but Kanhaiye and Vijay revealed the whereabouts of a hut that contained the stolen property, and the police took Sanjiv there. The police, along with Sanjiv, then went to Yusuf's house in Chembur, Bombay. Yusuf was not there, but his brother was, so the police beat Yusuf's brother and told him to bring Yusuf. His brother promised to send Yusuf to the police the next day.

The next morning, Sanjiv was produced before the metropolitan magistrate at Bhoiwada in the Dadar area of Bombay. The magistrate remanded him to fourteen days of police custody. The first two days, Sanjiv reported that he was not tortured, but following this, the torture began. He was made to stand with his hands tied to two stationary poles while police hit him with a wooden pole all over his body, including his face. Then his legs were tied and placed in a tire and he was beaten on the legs. After the beating, the police made him "exercise his legs" so that there would be no signs of torture.

Sanjiv was implicated in thirteen cases and remanded for two-and-a-half months to police custody, during which he was again badly beaten. He was then sent to Bombay Central Prison where he reported mistreatment at the hands of prison guards and inmates. While in prison, he came in contact with an NGO and now works for the same NGO.

The torture of children as a part of investigating crime is extremely common, as the following cases from Bangalore, Madras and New Delhi illustrate.

Kalasipalyam Police Station-Bangalore, July 7, 1995: Satish, sixteen, and Ayappa, twenty, were taken into custody by the Kalasipalyam Police Station for selling peanuts and cigarettes near a shop. This shop was robbed the next day, and the two boys were taken in for questioning. The boys pleaded that they knew nothing, but reported that the police beat them. After five days, an NGO secured their release. No charges were filed.

K-3 Police Station, Madras-September 1995: Shiva was sixteen and he had been on the street since 1992. He comes from the outskirts of Madras. He came to the street because he had no father, his mother was a drug-addict, and his twelve-year-old sister became pregnant. He decided to work as a ragpicker to help his sister and earns twenty to thirty rupees a day ($0.57 to $0.86).

He told Human Rights Watch that he was picked up by the police in September 1995 because they wanted information on a friend. He was taken to the K-3 police station and questioned about the whereabouts of his friend. The younger police officer started to beat him with a lathi, yelled at him, and asked him where his friend was. When he said he did not know and that he was an "NGO kid," an older officer said, "Do not beat NGO kids," and the younger officer stopped. After this they searched his rag-bag. They found plastic scraps, which they said may have been from stolen goods and kept him at the station overnight. They did not beat him any more.

Shiva said this about the police:

*Some are very good and they say don't sleep in the road, be careful. Some are very bad and they beat us brutally and take the money from our pockets.*

Vyalikaval Police Station-Bangalore, September 9, 1995: Murugan, aged thirteen, was apprehended by police from the Vyalikaval Police Station on September 9, 1995, at 6:00 a.m. on suspicion of stealing 4,000 rupees from the house where he worked as a car cleaner. The money had been lost one week earlier and the owner's wife had accused him of taking the money. Murugan had been waiting to collect his salary for the past seven months so that he could return home. A constable at the police station informed the NGO representative, who managed to get Murugan released, obtained his back pay for one-and-a-half years and sent him home to his family. Before being released, Murugan was again beaten by the police and given no food for the entire day.

Railway Protection Force Police Station-New Delhi Central Railway Station, New Delhi, December 1995: Girish was fourteen. He worked as an unlicensed porter at the New Delhi Railway Station. He was taken by the Railway Protection Force (RPF) to the RPF police station in late November-early December 1995. Girish told Human Rights Watch:

> *I was sleeping in the parking lot when the police came. They asked me who stole 10,000 rupees from a passenger. They said the passenger had left his suitcase at the station. I said I didn't do it. Then they asked me to name who did, but I didn't know anything. They took me to the station and two police started to beat me all over with lathis. They kept on saying, "Name the person," and beating me. I kept on telling them I didn't know anything, but they just beat me. They did this for about an hour and then they let me go. They didn't file any charges.*

## THE CONTROL OF CRIME AND DEVIANCE

Social control now refers to those institutions and mechanisms that define and respond to crime and/or deviance. Corresponding to the dominant theory groups in criminological sociology (the perspectives of crime causation, crime construction, and critical sociology), social control is now conceptualized as a functional response to crime, the societal reaction to deviance, or the reproduction of a social order beyond a mere focus on crime. First, from the perspective of crime causation theories, such as Edwin Sutherland's (1973) theory of differential association, social control is conceived as a dependent variable which functions, in response to crime, as a mechanism of redress. Crime takes centre stage in such a perspective as criminal behaviour, which needs to be causally explained, is observed to be sanctioned by the forces of social control in order to prevent the disintegration of society. This perspective of social control can theoretically rely on the functionalist sociology of Talcott Parsons (1951), who developed a perspective of social control as a corollary to a theory of deviance and crime. Whereas crime is seen as creating a tension in an otherwise stable system, social control is understood as a re-integrative attempt to stabilize the functioning of a society. The mechanisms of social control thus function to fulfil society's integrative needs in the societal community. It is to be noted that this conception of social control is not to be confused with the so-called social control theory of crime, which was developed by Travis Hirschi (1969). In Hirschi's theory, which is mostly an extension to Durkheim's theory of a society's need for integration and regulation (and what happens in the absence thereof), criminal behaviour is accounted for as the result of a weakening of the bonds with society. As a function of social control, conversely, society offers restraints on people's drives and desires.

Second, from the viewpoint of labeling or societal reaction theories, popularized by Howard S. Becker (1963) and Edwin Schur (1965), crime is viewed as a societal construction on the basis of a process of the criminalization of deviance. Whereas an act of deviance is seen as motivated by an actor, its subsequent criminalization is conceived as a function of the societal conditions that define and respond to deviance. Social control is thus constructive of crime through a process of labeling.

From the perspective of crime construction theories, social control mechanisms are not a functional response to crime, but instead determine crime. This occurs through two processes of criminalization: through primary criminalization, some acts are defined as criminal; and through secondary criminalization, this definition is applied to specific acts. The formal treatment of crime through the processes of social control, furthermore, is typically observed to not take into account the needs and motives of the deviant actor, but instead imposes a system of control that serves societal goals of retribution and punishment.

Third and finally, from the viewpoint of sociological conflict theory, the interactionist focus of labeling theory is transcended with a structural perspective that situates the labelling processes of social control within the broader society in which they take place (e.g., Quinney 1973). Instead of analysing the interactionist order of rule-violator and rule-enforcer, a critical sociology focuses on social control in terms of the historically grown socio-economic conditions of society and its mechanisms and institutions that are mobilized to maintain order. Marxist perspectives, for instance, focus on the norms of enforcement as representing the interests of a class, the elite, which try to have their interests accepted as general norms. Non-Marxist critical perspectives more broadly focus on a variety of groups and inequalities (classism, racism, sexism, ageism).

To end this review of the concept of social control in relation to crime and deviance, it is to be noted that there have been continued attempts in modern sociology to redefine social control broadly, specifically in the works of Morris Janowitz and Jack Gibbs. Morris Janowitz (1975, 1978) uses the concept of social control to denote a society's capacity to regulate itself within a moral framework that transcends self-interest. He argues that social control has been drastically weakened in advanced industrial societies, and wonders why this is the case and how social control can be re-strengthened.

Jack Gibbs (1989, 1994) has developed a scientific theory with a high degree of predictive power (accuracy, testability, scope, range, intensity, discriminatory power, and parsimony). He defines (attempted) control as "overt behaviour by a human in the belief that (1) the behaviour increases or decreases the probability of some subsequent condition and (2) the increase or decrease is desirable" (Gibbs 1994:27). Distinguished on the basis of the target of control are inanimate, biotic, and human control, pertaining to control over objects, nonhuman organisms, and humans, respectively. Control over human behaviour comprises self-control, proximate, sequential, and social control, relative to how many and how other humans are involved.

Turning to contemporary theories of social control in relation to deviance, at least two very different strands of theorizing can be highlighted: the general theory of social control developed by Donald Black; and the revisionist perspective of social control that builds on the work of Foucault.

## A General Theory of Social Control

Originally conceived by Donald Black, the general theory of social control aims to provide a scientific theory of social control that is based on the principles of natural science (Black 1976, 1997). Black is interested in formulating a general theory of law and social control that can account for empirical variation irrespective of any value judgments or policy claims. Conceiving of law as governmental social control, Black's pure sociology is oriented at a general theory of all forms of social control, defined as the handling of right and wrong by defining and responding to deviant behaviour. The epistemological orientation that underlies Black's pure sociology is scientific in its ambition to formulate a general theory, whereby the ordering of variation in empirical reality is seen as the goal of theory. The paradigmatic framework in which Black's theory is situated is distinctly sociological. Rejecting teleological and anthropocentric premises that take into account, respectively, normative and subjective dimensions, Black's approach is radically anti psychological in developing a multi dimensional perspective of social life, including law, as a function of structural characteristics of social space.

On the basis of this paradigm, Black has developed a number of propositions on the behaviour of law and other forms of social control. Social control is generally conceived of as a reality that appears in variable forms of quantity and style. The quantity of social control refers to the amount of social control that is available, for instance whether or not a particular kind of human conduct is regulated by law and whether or not a legal sanction is applied. The styles of social control can be of various kinds, such as penal, compensatory, therapeutic, or conciliatory. In seeking to account for variations in social control in terms of quantity and style, Black studies the geometry of social control on the basis of variations in social space in terms of such characteristics as stratification, differentiation, integration, and culture. For instance, stratification refers to the vertical structure of society in terms of the inequality of wealth. Vertical space can be high or low in terms of position or downward or upward in direction. Morphology refers to the horizontal aspect of society, including the division of labour (differentiation) and the relative degrees of intimacy and distance (integration). Culture is the symbolic dimension of social life, including expressed ideas about truth, beauty, and ethics, such as in science, art, and religion. Culturally, societies and social groups can vary from being closely related to extremely distant.

On the basis of the suggested model, Black develops various testable propositions. Among them is the thesis that social control varies directly with stratification: societies with higher degrees of stratification have more social control. Law and social control also vary directly with culture: simpler societies have less law and social control than more differentiated societies. On the basis of such propositions, pure sociology seeks to explain and predict the behaviour of social life in terms of social space in value-neutral terms. Irrespective of the intrinsic merits of Black's approach, it is striking that the sharpness of his formulations has enabled much theoretical debate and empirical research.

## GENERAL THEORY OF CRIME

In the relatively short period of time since its publication, *A General Theory of Crime* (1990) has seemed to attract an impressive amount of attention from criminologists. Travis Hirschi, in collaboration with Michael Gottfredson, moved away from his classic social bonding formulation of control theory and developed *A General Theory of Crime* (1990). In Hirschi's original social bonding theory (1969), he emphasized the importance of "indirect control"—which allows parents to have a "psychological presence" when youths are not under their surveillance, additionally, this theory contained four elements of control: attachment, commitment, involvement, and beliefs. However, Gottfredson and Hirschi argue "direct control" is the key to the most effective parenting. For this reason, they proposed a theory of crime based solely on one type of control alone—self-control. They offer self-control theory as a generalized theory that explains all individual differences in the "propensity" to refrain from or to commit crime, which they point out includes all acts of crime and deviance at all ages, and circumstances (Akers and Sellers, 2004:122). Gottfredson and Hirschi begin with the observation that: Individual differences in the tendency to commit criminal acts... *remain reasonably stable with change in the social location of individuals and change in their knowledge of the operation of sanction systems*. This is the problem of self-control, the differential tendency of people to avoid criminal acts whatever the circumstances in which they find themselves. Since this difference among people has attracted a wide variety of names, we begin by arguing the merits of the concept of self-control (Gottfredson and Hirschi, 1990: 87).

First, it should be noted that Gottfredson and Hirschi differentiate between "criminality," which is the propensity to offend, and "crime," which is an actual event in which a law is broken. They recognize that a propensity cannot be acted on unless the opportunity to do so exists. Consequently, they see crime as a by-product of people with low self-control, who have high criminogenic propensities, coming into contact with illegal opportunities. Still, given the most offenses are easy to commit and opportunities for crime are constantly available, over time people with low self-control inevitably will become deeply involved in criminal behaviour. That is, self-control, not opportunities, will be the primary determinant of people's involvement in crime across their life course (Agnew and Cullen, 1999:175).

It is seen that low self-control develops early in life and remains stable into and through adulthood. Gottfredson and Hirschi trace the root cause of poor self-control to inadequate childrearing practices. Parents or guardians who refuse or who are unable to monitor a child's behaviour, who do not recognize deviant behaviour when it occurs, and who do not punish that behaviour will produce children who lack self-control. As Dennis Giever explains, "children who are not attached to their parents, who are poorly supervised, and whose parents are criminal or deviant themselves are the most likely to develop poor self control" (1995). Consequently, a lack of self-control occurs naturally in a child when steps are not taken to stop its development.

Gottfredson and Hirschi suggest that high self-control effectively reduces the possibility of crime –namely, those possessing it will be substantially less likely at all periods of life to engage in criminal acts (Gottfredson and Hirschi, 1990:89). In contrast, the lower a person's self-control, the higher his or her involvement in criminal behaviour and in acts analogous to crime. These individuals often have a tendency to respond to tangible stimuli in their immediate environment, specifically, having a "here and now" orientation. As Gottfredson and Hirschi most notably state, they derive satisfaction from "money without work, sex without courtship, revenge without court delays" (1990:89). People lacking self-control also tend to lack diligence, tenacity, or persistence in a course of action. To these individuals criminal acts tend to be exciting, risky, or thrilling and they maintain an adventurous point of view. In addition, these crimes provide few or meager long-term benefits, require little skill of planning, and often result in pain or discomfort for the victim. "In general, there is a fairly consistent support for Gottfredson and Hirschi's theoretical predictions—a fact that ensures that their self-control theory will remain an important theoretical perspective in the time ahead".

Gottfredson and Hirschi stress that there is great versatility in the types of crime and analogous behaviour committed by persons with low self-control. Self-control, according to the theory, accounts for all variations by sex, culture, age, and circumstances and "explains all crime, at all times, and, for that matter many forms of behaviour that are not sanctioned by the state", and is "for all intents and purposes, *the* individual-level cause of crime". This is a bold claim made by both Gottfredson and Hirschi in which they attempt to support by reviewing known official and unofficial distribution and correlates of crime and delinquency, interpreting them as consistent with the concept of self-control.

Despite much criticism, the body of empirical tests of the general theory of crime has been fairly consistent in revealing a link between self-control and crime (Pratt & Cullen, 2000). Grasmick et al. (1993) provides influential research to the area of control theory using a community sample of 395 adults from Oklahoma City to probe six different dimensions of self-control derived from Gottfredson and Hirschi's theory. The six different dimensions included were: (1) impulsivity; (2) a preference for simple tasks; (3) risk-seeking; (4) physicality; (5) self-centeredness; and (6) a bad temper. Of 24 items that they proposed to tap the self-control concept, the principle components analysis suggested that 23 clung together to form a reliable and unidimensional self-control scale (Crombach's Alpha=.81). Similarly, Alexander Vazsonyi and his associates found that the self-control scale developed by Grasmick et al. (1993) had a similar predictive power when they analysed self-control and deviant behaviour with samples drawn from a number of different countries (Hungary, Switzerland, the Netherlands, the United States, and Japan). They found that low self-control is significantly related to antisocial behaviour and that the association can be seen regardless of cultures or national settings (Vazsonyi et al., 2001). Additionally, Pratt and Cullen (2000) conducted a meta-analysis of 21 cross-sectional and longitudinal studies directly testing the relationship between low self-control and crime, some of which used behavioural and others of which used attitudinal measures of self-control. On average, the self-control variables explained 19 percent of the variance in delinquent and criminal behaviour, which were consistent in the expected direction. "Low self-control must be considered an important predictor of criminal behaviour," but studies do not support the argument that self-control is the sole cause of crime or that the "perspective can claim the exalted status of being the general theory of crime".

However, some critics argue that the theory is tautological or involves a certain degree of circular reasoning. For example, one begins with the definition of low self-control as the failure to refrain from crime and then proposes low self-control as a cause of law violation, thus one's proposition is tautological. Given this definition of low self-control, the very thing it is hypothesized to explain defines it; hence, the proposition can never be proven false. A general theory of crime hypothesizes that low self-control is the cause of the propensity toward criminal behaviour. Specifically, in regard to the theory's testability, Gottfredson and Hirschi do not define self-control separately from this propensity. Incidentally, "they use 'low self-control' or 'high self-control' simply as labels for this differential propensity to commit or refrain from crime. They do not identify operational measures of low self-control as separate from the very tendency to commit crime that low self-control is supposed to explain. As a result, the propensity toward crime and low self-control appear to be one and the same".

Akers (1991) illustrates that to avoid this tautological problem, conceptual definitions or operational measures of self-control must be developed that are separate from measures of criminal behaviour or propensity toward crime. However, additionally the theory of low self-control is seen as being logically consistent, parsimonious (conciseness and abstractness of a set of concepts and propositions), and having a wide scope. It has generated enormous interest and attention in the field of criminology and some contend that it may supersede social bonding as the principal control theory. Thus far, the tautology issue has not been resolved, but some research is propelling toward that direction by measuring self-control independently of measures of crime propensity. While some research reports continually contradict the theory, and the broad claims of being the explanation of criminal and deviant behaviour cannot be sustained, on balance the empirical evidence supports the theory.

Perrone, Sullivan, Pratt, and Margaryan (2004) conducted an empirical test of the general theory of crime in which they examined the relationship of parental efficacy, self-control, and delinquency. The data was obtained from a nationally representative sample of adolescents (the National Longitudinal Study of Adolescent Health). This study aims to address two important questions, 1) whether parental efficacy is a significant predictor of youths' levels of self-control, and 2) whether self-control mediates the relationship between parental efficacy and delinquency. In regard to the second question at hand, "the current analysis has important implications for the comparative validity of self-control theory versus social learning and developmental/life-course perspectives—both of which are capable of providing a more convincing explanation of the existence of direct and indirect (through self-control) effects of parental efficacy on delinquent behaviour".

Conducted by researchers at the Carolina Population Centre, the data for this research was drawn from the first wave of the Add Health study (the

National Longitudinal Study of Adolescent Health). By use of a stratified random sample of all high schools in the United States, initially, researchers chose 80 high schools from clusters based on several characteristics: region, urbanization, school size and type (public or private), race, and grade span. More than 70% of the originally sampled high schools agreed to participate in the study in which adolescents from grades 7 to 12 were randomly selected from the school provided rosters. In addition, data from the wave I in-home sample was used by means of qualitative study. The in-home component of this study was used because it includes data for adolescents and their parents assessing a variety of different measures. A total random sample of 15,243 adolescents was selected for the in-home portion of the Add Health study (Resnick et al., 1997).

When conducting the in-home interview portion of the study, preference was given to resident mothers or other female caretakers, as previous studies found that mothers were more knowledgeable about their children's health and behaviour than fathers (Bearman et al., 1997). The following measures were assessed for both adolescents and parents:

*Delinquency Scale.* Which consists of 6-items created to measure delinquent behaviour on the part of adolescents (á =.66, M= 2.76, SD = 1.776). "The Delinquency Scale included questions on lifetime use of cigarettes and alcohol, and the number of times the adolescent smoked marijuana. Furthermore, the adolescents were asked the number of times they 'lied to their parents,' 'engaged in a serious physical fight,' and 'behaved disorderly in a public place' within the past year. A score of 1 was given if the individual responded yes to this behaviour and a score of 0 was given if he/she did not. The scale is composed of the average response across the six items".

*Self-control.* "Gottfredson and Hirschi (1990) described six dimensions of self-control: impulsivity, a preference for simple tasks, the favoring of physical over mental activities, self-centeredness, and a temper component. However, their unidimensional factor contains five items that tap into five of Gottfredson and Hirschi's six self-control dimensions.

Given the potentially tautological nature of behavioural measures of self-control, we combined attitudinal and behavioural measures. For instance, adolescents were asked to respond to the following statements: Have they 'had problems keeping their mind on what they were doing,' have they 'had trouble getting their homework done,' and have they had difficulty 'paying attention in school.'

These questions tap into the simple tasks, physical activities, and impulsivity components of self-control. Respondents were also asked whether they had trouble getting along with their teachers, which captures self-control's temper dimension. Another question tapped the self-centeredness of the respondents by asking them to respond to the following statement: 'you feel you are doing everything just about right.' Because a response of 'yes' to these items indicated a low level of self-control, higher values on our self-control measure indicate lower levels of self-control.

*Parental efficacy.* "Our unidimensional parental efficacy factor is made up of four items that address a mother's attachment to her child (e.g., 'Do you get along with your child?') As well as the mother's effectiveness in recognizing problematic behaviour and responding to this behaviour. High scores on this scale indicate higher levels of parental efficacy".

In addition, control variables were also taken into account by researchers. They controlled for a host of additional demographic and social characteristics for each respondent to isolate the effects of parental efficacy on self-control and the effects of parental efficacy and self-control on delinquency.

As stated above, the current study had two purposes: to examine (1) whether parental efficacy is a significant predictor of youths' levels of self-control and (2) whether self-control mediates the relationship between parental efficacy and delinquency. In regard to the first question, parental efficacy was found to be a significant predictor of youths' levels of self-control. Furthermore, these findings are consistent with propositions set forth by Gottfredson and Hirschi (1990) regarding the development of self-control in children. Secondly, assessing whether self-control mediates the relationship between parental efficacy and delinquency, findings indicate that, at best, self-control only partially mediates this relationship. Additionally, this finding is in direct opposition with Gottfredson and Hirschi's (1990) proposition that self-control should fully mediate the parental efficacy-delinquency relationship.

Accordingly, the central purpose of the current study was to re-examine the dynamics of parental efficacy, self-control, and delinquency using a large, nationally representative sample of youths with more temporally proximate measures of parenting. In doing so, the results of their analysis yielded four major conclusions. First, which was consistent with Gottfredson and Hirschi's (1990) framework, parental efficacy is a major precondition for self-control in youngsters. Secondly, "although the parental efficacy-self-control link was quite robust, their measures of race and family structure (along with age and sex) were significantly related to self-control. These findings highlight two important issues relevant to criminological theory and research. First, this analysis indicates the importance of family context, not simply patterns of parental monitoring and supervision, to the explanation of delinquency".

Third, and related, their results indicate that the dynamics of race and self-control may be much more intricate than indicated by previous research. "Specifically, the over-sampled middle-and upper-middle-class Blacks in the sample exhibited relatively high levels of self-control. Finally, their fourth major conclusion was that the ability of self-control to mediate the relationship between parental efficacy and delinquency was, at best, limited". "Thus, when taken together it appears that parental efficacy affects delinquency in ways that are not easily explained by Gottfredson and Hirschi's (1990) theory. These findings also have important implications for continued theoretical development and integration in criminology".

## LEARNING THEORIES OF CRIME

Some theories in criminology believe that criminality is a function of individual socialization, how individuals have been influenced by their experiences or relationships with family relationships, peer groups, teachers, church, authority figures, and other agents of socialization. These are called learning theories, and specifically social learning theories, because criminology never really embraced the psychological determinism inherent in most learning psychologies. They are also less concerned for the content of what is learned (like cultural deviance theories), and more concerned with explaining the social process by which anyone, regardless of race, class, or gender, would have the potential to become a criminal. Social Learning, Control, and Labeling theories are all examples of social process theories.

Learning is defined as habits and knowledge that develop as a result of experiences with the environment, as opposed to instincts, drives, reflexes, and genetic predispositions. Associationism (developed by Aristotle, Hobbes, Locke, and Hume) is the oldest learning theory. It is based on the idea that the mind organizes sensory experiences in some way, and is called cognitive psychology today. Behaviorism (developed by Pavlov and Skinner) is the second oldest learning theory. It is based on the idea that the mind requires a physical response by the body in order to organize sensory associations. There are two types of learning in behavioral psychology: classical conditioning (where stimuli produce a given response without prior training); and operant conditioning (where rewards and punishments are used to reinforce given responses). Examples of operant conditioning include verbal behaviour, sexual behaviour, driving a car, writing a paper, wearing clothing, or living in a house. *Most social behaviour is of an operant nature*. Imitation (sometimes called contagion) is the oldest social learning theory, and derives from the work of Tarde (1843-1904), a sociologist who said crime begins as fashion and later becomes a custom.

The Social learning theory that has had the most impact on criminology is associated with the work of Bandura (1969), a psychologist who formulated the principles of "stimulus control" (stimulus-to-stimulus reinforcement rather than stimulus-behaviour reinforcement), outlined the stages of "modeling" (attend, retain, rehearse, perform), and pioneered the field of "vicarious learning" (media influences, for example).

Of these many contributions, the one about stimulus-to-stimulus chains of learning is the most important since it does away with the need for extrinsic rewards and punishments, arguing that observational learning can take place without them.

Bandura's ideas about role modeling resonated well with criminology because since the 1930s, criminology had a similar theory (differential association). Julian Rotter was also another psychologist who had an enormous impact on social learning theory in criminology.

## Sutherland's Differential Association Theory

Sutherland (1883-1950) is called the father of American criminology. In 1924, he wrote a book called *Criminology*, the first fully sociological textbook in the field. He first put forth his theory in the second edition of 1934, revised it again in 1939, and the theory has remained unchanged since the fourth edition of 1947. When Sutherland died in 1950, Donald Cressey continued to popularize the theory.

It's called Differential Association (DA) theory, and Sutherland devised it because his study of white collar crime (a field he pioneered) and professional theft led him to believe that there were social learning processes that could turn anyone into a criminal, anytime, anywhere. Let's look at the 9 points of DA theory:

1. Criminal behaviour is learned....
2. Criminal behaviour is learned in interaction with others in a process of communication....
3. Learning criminal behaviour occurs within primary groups (family, friends, peers, their most intimate, personal companions)
4. Learning criminal behaviour involves learning the techniques, motives, drives, rationalizations, and attitudes....
5. The specific direction of motives and attitudes is learned from definitions of the legal codes as favourable or unfavorable....
6. A person becomes a criminal when there is an excess of definitions favourable to violation of law over definitions unfavorable to violation of law.... (this is the principle of differential association)
7. Differential associations vary in frequency, duration, priority, and intensity (frequent contacts, long contacts, age at first contact, important or prestigious contacts)
8. The process of learning criminal behaviour involves all the mechanisms involved in any other learning....

9. Although criminal behaviour is an expression of general needs and attitudes, criminal behaviour and motives are not explained nor excused by the same needs and attitudes (criminals must be differentiated from noncriminals)

Sutherland's theory is tested mainly on juveniles because delinquency is largely a group crime. The principle research problem is determining which comes first, the delinquency or delinquent friends. It may be that "birds of a feather flock together" and delinquency "causes" delinquent friends, but delinquent friends do not cause delinquency. There are plenty of negative test cases showing that not everyone who associates with criminals becomes a criminal.

*The theory doesn't even try to explain where the first, original unfavorable definition comes from.* There's some research showing that more recent than earlier delinquent friends have more influence (Warr 1993), and other research (Short 1957;Jensen 1972) has produced mixed findings. Another problem is the complex modeling involved (Matsueda 1988) to specify the causal structure in Sutherland's theory, especially the ratio of definitions favourable and unfavorable to violating the law, as the following examples indicate:

| Definitions favourable: | Definitions unfavorable: |
|---|---|
| Fair play | Cheating and shortcuts are OK |
| Forgive and forget | I don't get mad, I get even |
| Good always wins | Sometimes evil wins |
| Give others a chance | Take advantage of suckers |

Despite these criticisms, DA theory applies to most types of crime (upper world and underworld; juvenile and adult; it doesn't explain spontaneous, wanton acts). It sensitizes criminologists to the role of ideas, not social conditions, as influences on criminal behaviour.

## Akers' Differential Reinforcement Theory

In the late 1960s, new social learning theories were developed which dropped Sutherland's point that learning criminal behaviour takes place in primary groups (Burgess & Akers 1968). Differential reinforcement theories, as they were called, tried to incorporate the psychological principles of operant conditioning, and held that even nonsocial situations (such as the physical effects of drug use) could reinforce learning criminal behaviour. In sociology, there's also a group of theories called differential identification, which deal with imaginary, or comparative reference groups as influential in perceiving deviance as respectable.

According to Akers (1985), people are first indoctrinated into deviant behaviour by differential association with deviant peers. Then, through differential reinforcement, they learn how to reap rewards and avoid punishment by reference to the actual or anticipated consequences of given behaviours. These consequences are the social and nonsocial reinforcements that provide a support system for those with criminal careers or persistent criminality. Structural conditions affect a person's differential reinforcements. Criminal knowledge is gained through reflection over past experience. Potential offenders consider the outcomes of their past experiences, anticipate future rewards and punishments, and then decide which acts will be profitable and which ones will be dangerous. To get to the point where criminal behaviour is activated by discriminative cues (norms), the whole process follows seven stages:

1. Criminal behaviour is learned through conditioning or imitation.
2. Criminal behaviour is learned both in nonsocial reinforcing situations or nonsocial discriminative situations and thru social interaction.
3. The principal components of learning occur in groups.
4. Learning depends on available reinforcement contingencies.
5. The type and frequency of learning depends on the norms by which these reinforcers are applied.
6. Criminal behaviour is a function of norms which are discriminative for criminal behaviour.
7. The strength of criminal behaviour depends upon its reinforcement.

This theory has been tested fairly successfully on teenage deviance such as tobacco, drug, and alcohol use (Akers et al. 1979). It tends to explain 50-60% of the differences between users and abstainers using measurement items like: perceptions of actual or anticipated praise or absence of praise; and the total good things felt from deviance minus the total bad things felt. Differential reinforcement theory tends to fit well with rational choice theory because they both explain the decision making process involved in developing the motivation, attitudes, and techniques necessary to commit crime. It can explain solitary offending (since learning is sometimes solitary). Like all learning theories, it claims to be a general theory of crime.

## Jeffery's Differential Reinforcement Theory

Not to be overlooked, Jeffery's (1965) theory of differential reinforcement is based on the ideas of conditioning history, deprivation, satiation, the proceeds of crime being reinforcing in themselves, and the absence of punishment. This last variable (absence of punishment, or not getting caught) makes the theory tend to resemble a social control theory of crime (Conger 1976). Jeffery takes for granted the existence of constant stimuli in the environment, but argues that the key variables of deprivation and satiation are what make these stimuli reinforcing or not. A person deprived of something will respond to a stimulus in a much different way than a person satiated with something. Crimes

# Crime and Forensic Science

against property produce things, like money or cars, which are themselves powerful reinforcers. Crimes against persons most often involve what Jeffery calls negative reinforcement, the removal of an aversive stimulus (such as when a man kills his unfaithful wife). Drug offenses are reinforced by biochemical changes in the human body.

A criminal act may lead to reinforcement, but it may also lead to punishment. Punishment is an important variable in Jeffery's theory because, essentially, he argues that crime occurs because criminal acts in the past, for a particular actor, have not been punished enough. Criminals have different conditioning histories, but it is possible for a person living in a criminal environment to not become a criminal and for criminals to be found in noncriminal environments. Associations with others do not matter. *The behaviour is shaped by discriminative stimuli rather than reinforcing stimuli.*

Jeffery's theory is compatible with the classical school of criminology in that the perceived certainty of punishment, not its severity, is what deters people from criminal acts. He's also concerned with sloppy administration of the criminal justice system, in the sense that it's administration produces avoidance and escape responses rather than aversive consequences. Hence, a criminal doesn't refrain from crime, but from apprehension, detection, and conviction (by not leaving fingerprints, hiring a good lawyers, pleading insanity, etc.).

## Matza's Neutralization Theory

Neutralization theory (Sykes & Matza 1957) holds that people learn the values, attitudes, and techniques of criminal behaviour through subterranean values, which exist side by side with conventional values. Few people are "all good" or "all bad." Matza argues that most criminals are not involved in crime all the time. They drift from one behaviour to another, sometimes deviant, sometimes conventional. Sykes and Matza state that criminals frequently admire honest, law abiding persons, and therefore are not immune from guilt and the demands of conformity. They have to use excuses that allow them to drift into crime. In this way, neutralization theory is a control theory if the excuses are seen as post-hoc rationalizations. If they exist in a subterranean fashion (as originally thought), then it's a learning theory. They identified the following techniques of neutralization that criminals use to escape from the demands of conventionalism:

1. Denial of responsibility — It's not my fault; I didn't have a choice
2. Denial of injury — It's no big deal; They have too much money
3. Denial of victim — They had it coming; They had a bad attitude
4. Condemnation of the condemners — Everybody does it; Why me?
5. Appeal to higher loyalties — Only cowards back down; protecting

Neutralization theory can explain the behaviour of occasional criminals, like shoplifters or poachers. Further, specific types of crime may be related to specific types of excuses. Learning theories, in general, have the capacity to explain a wide assortment of criminal activities. Agnew (1994), for example, applied neutralization theory to violent crime, and Minor (1980) has come up with several extended neutralization techniques. Some criminologists think of neutralization theory as a CONTROL theory while others believe it fits into a symbolic interactionist, or LABELING, perspective which focuses upon identity formation processes.

## CLASS AND CRIME

The longstanding controversy over the importance of social class in the production of criminal conduct is often an argument over the meaning of class and the measurement of crime. Criminal conduct is far from a unitary phenomenon. In general, for a crime to be committed, there must be some intentional conduct that is prohibited by a criminal law. Occasionally, the law may require specific conduct such as filing a tax return. Under these circumstances, a lawmaking body can create a link between class and crime simply by making rules designed to control the conduct of the rich or the poor. If the legislature creates a law making it a crime to be found in public without money or a permanent address, they will have created a link between poverty and crime. If they make it a crime to engage in "insider trading" on the stock market, they will have created a crime that is almost certain to involve those with access to management decisions that might change stock prices. This kind of law would create a link between wealth and crime.

### Definition Of Crime

Although official definitions of crime are legislative, in practice crime is defined by administrative policies and enforcement practices. While most crime is some form of theft or assault and most of it results in physical harm or property loss for individuals, there are crimes where no loss of property is involved and no injury is inflicted on others. Enforcement policies and practices will determine who is arrested for such crimes. The areas in which these offenses are perpetrated, as well as the prior income and employment status of prison and jail inmates suggest that drug laws and laws against gambling and prostitution have generally worked against the poor more than they have against the rich.

Those who study crime and delinquency also define crime. The definition of crime was greatly expanded when criminologists began asking people to report their own illegal or improper behaviour. In some of the early self-report studies, conduct that is only illegal when minors do it was defined as criminal

(Nye and Short). In some self-report studies conduct was defined as delinquent even when it was so common than almost everyone could be classified as delinquent. At the other extreme, criminologists have classified some conduct as criminal that does not violate existing law. These writers believe that all forms of economic exploitation, racial discrimination, or creation of unsafe or unhealthy work environments are harmful and should be made criminal. Because they define such conduct as criminal, they argue that crime is evenly distributed across class levels or that it is linked to upper class status (Pepinsky and Jesilow).

## Measuring Crime

Some measures of crime are based on police, court, correctional, or official survey reports. These efforts produce information on victims and offenders. Reports of offenses known to the police and victimization survey results provide victim-based information. However, such victim information is sometimes used to infer offender characteristics. On occasion, victim-based measures are simply treated as if the offender-victim distinction is unimportant. That is, the focus on victims in such studies is never mentioned. Occasionally, offender information, such as that provided by the Supplementary Homicide Reports (SHR) program or by police reports of arrests, is used to modify victim information.

A few studies have used arrest data in combination with offenses known to the police to create race-specific offence rates (Sampson; Ousey). More often, offender information is used to look at offender characteristics or the relationship between victims and offenders (Chilton and Jarvis). It is sometimes used to compute rates for studies that examine the relationship of offence rates to other economic and social characteristics of urban areas.

A different set of crime measures are created when interviews or questionnaires are used to ask people about crimes they have committed. Those asked about their criminal conduct can be juveniles or adults, male or female. They may live in the same community or be part of a national sample. The measures of crime used in such studies vary widely. Respondents may be asked to select, from a list, offenses they have committed at some point in their lives or at some time during the last year. They may or may not be asked about the frequency with which they have engaged in such conduct. The acts presented range from very minor offenses, or offenses that are only illegal for children, to very serious offenses. Measures of crime are sometimes created by counting the number of different types of crime reported and sometimes by using the frequency of crimes reported or by counting specific offenses such as assault or burglary.

## Definition Of Class

In addition to issues of the definition and measurement of crime, disagreements about the meaning and measurement of social class make it difficult to conclude whether or not class is linked to crime. Looking at social class categories as essentially a matter of differences in wealth and income, we can say in a general way that those who own a great deal of property and have high incomes are rich or upper class; those who own little or nothing and have low incomes are poor or lower class. Beyond this general notion the issue is quickly complicated. No commonly accepted set of classes exists. And a wide variety of gradational scales designed to measure social class have been developed. Self-report studies generally use reports of parent's occupation to create social class scores. At least one self-report study of adults asked for work information and used it to assign each respondent to a specific social class depending on his or her business ownership and employee or employer status.

Studies of geographic distribution are more likely to infer the social class of an area based on measures that reflect the income and assets of those living in the area. Measures often used are the median income of the residents of each area, the proportion of home ownership, the median value of homes, median rent, the proportion of the population in poverty, median education, and the prevalence of dilapidated housing. Variations on these indications of area wealth and deprivation are sometimes used. Results vary according to the measures used and their construction and, more often, according to the size of the areas used—census tracts, cities, Metropolitan Statistical Areas (MSA), or states. An additional complication in discussions of the social class of geographic areas arises because it is possible to see people as rich or poor in either an absolute or relative sense. This has produced studies of inequality and crime in addition to, and sometimes instead of, poverty and crime. In such an approach the emphasis is on the gap between those with high incomes and those with low incomes.

## Early Work

For the first half of the twentieth century, the question of the link between class and crime was examined in three basic ways. First, investigators looked at the impact of economic conditions on crime rates, asking if crime increases with an economic downturn. A basic assumption in this approach was that poor economic conditions are harder on the poor than the middle class and that this produces increased crime. A second approach examined the social class of prisoners or others formally identified as offenders to ask about the social class backgrounds of people convicted of crime. Generally, convicts were and are poor. In a third approach, crime rates for specific geographic areas were compared with a set of social and economic characteristics of the areas. These studies asked if areas with indications of high poverty rates and low social class were also areas with high crime rates. In general the answers to this question were yes. All three of these approaches probably influenced the development of theories either attempting to explain the reasons for the class-crime relationship or assuming such a relationship.

Some of the earliest empirical efforts to study class and crime used measures of the general economic conditions of regions of a country in combination with official crime rates for the regions to ask if poor economic conditions were associated with high crime rates (Bonger). Although those carrying out these studies often found that poor regions had high crime rates, they also found poor regions in which the crime rates were low. This led Bonger to conclude that the gap in income and wealth between the rich and poor might be more important than the overall poverty or affluence of an area. When similar studies were done for areas within cities in the early decades of the twentieth century, most suggested a clear link between crime or delinquency rates and the social and economic characteristics of urban areas. By the 1940s there was general agreement that both property crimes and crimes of violence were higher in areas with low average incomes, high transiency, low educational achievement, and high unemployment (Shaw and McKay).

In addition, examinations of the characteristics of prisoners during the first half of the twentieth century indicated that a disproportionate percentage were poor, uneducated, and unemployed before incarceration (Glueck and Glueck). In general, most of these early examinations suggested there was a class-crime link. Moreover, since the relationship could be interpreted as showing that poverty and unemployment produced much ordinary crime, the findings at the early studies were consistent with conclusions reached by a number of philosophers and social thinkers.

## SHIFTS IN FOCUS

In the 1940s and 1950s there was a shift in focus in criminology. The first aspect of the shift came when Edwin Sutherland introduced the notion of "white collar crime" to call attention to offenses committed by high status people in conjunction with their occupations. As he saw it, this occurred in two ways. Some high status individuals, acting alone, engaged in large-scale theft by embezzlement or fraud. In addition, groups of high status individuals, acting in concert, engaged in what he called "corporate crime." This frequently involved corporate efforts to reduce competition through some form of price-fixing. It sometimes involved the intentional manufacture and sale of toxic or dangerous products. Thus, "white collar crime" shifted the focus from the poor to the wealthy and is sometimes used to argue against the notion that poverty increases most forms of crime.

A second shift in focus came at about the same time when some criminologists fixed their attention on young people and on middle-class delinquency. Two research procedures were important in this shift. One was the development of self-reported crime studies (Nye and Short). The other was the use of techniques that required researchers to spend time with and observe the actions of middle-class young people. Both of these developments led investigators to conclude that there was a great deal of unreported criminal and delinquent conduct committed by middle-class children. Interest in the observation of middle-class children waned but interest in confessional studies was strong in the 1960s and 1970s and remained strong through the end of the century.

Almost all of the self-report studies used samples of young people in school who were assured of anonymity. Some national samples of minors were selected along with a few studies of adults. In some studies, the children were interviewed more than once and some were followed into adulthood. Most of these studies found weak or nonexistent links between social class and juvenile delinquency or crime. However, some studies using national samples to measure the frequency of self-reported delinquency found that lower-class youth reported nearly four times as many offenses as middle-class youth and one and one-half times as many as working-class youth (Elliott and Ageton).

In trying to reconcile the conflicting results of a number of individual-level confessional studies with those comparing area characteristics with area crime rates, some questioned the accuracy, representativeness, and scope of the surveys. Others played down or ignored the problems presented by the survey approach and concluded that the impact of social class on crime was a myth.

In 1979, John Braithwaite published a careful review of a large number of area and confessional studies and a balanced discussion of the advantages and limitations of each. After reviewing studies carried out through the mid-1970s, he concluded that lower-class children and adults commit the types of crime handled by the police at higher rates than middle-class children and adults. On the "myth" of the class-crime relationship, he warns us "be wary of reviews that pretend to be exhaustive but are in fact selective".

# Bibliography

Ahuja, Vijay: *Network and Internet Security*. Academic Press, Inc.: Boston, 1996.

Allport, G. W. : *Becoming: Basic Considerations for a Psychology of Personality,* Yale Univ. Press, New Haven, CT. 1955.

Amoroso, E. : *Fundamentals of Computer Security Technology,* Prentice Hall, Englewood Cliffs, 1994.

BloomBecker, Buck: *Spectacular Computer Crimes*. Dow Jones-Irwin: Homewood, IL, 1990.

Chambliss, W.: *Sociological Readings in the Conflict Perspective,* Mass.: Addison-Wesley, 1973.

Charlotte Waelde, *Law and the Internet: Regulating Cyberspace*. Hart Pub, 1997.

Chuck Easttom: *System Forensics, Investigation, and Response*. Jones & Bartlett. 2013.

Clausewitz, C. V. : *Principles of War*, Minneola, New York, Dover Publication, 2003.

Felson, Marcus: *Crime and Everyday Life, Insights and Implications for Society*, Thousand Oaks, California, Pine Forge Press, 1994.

Ferguson, M. *The aquarian conspiracy,* Jeremy P. Tarcher, Los Angeles, 1980.

Frederick B.: *Protection and Security on the Information Superhighway*. John Wiley & Sons, Inc.: New York, 1995.

Georgi, A. : Psychology as a Human Science, Harper & Row, New York, 1970.

Goitein, L. : *The Importance of the Book of Job for Analytic Thought*, American Imago, 1954.

Gottfredson, Michael R., and Travis Hirschi: *A General Theory of Crime*, Berkeley, California, Stanford University Press, 1990.

Gupta, A., & Laliberte, S. : *Security by Example Defend I.T.* New York, Addison Wesley, 2004.

Holland, N. : *The Dynamics of Literary Response*, New York, Oxford University Press, 1968.

Horace R. Cayton: *Black Metropolis, A Study of Negro Life in a Northern City*, New York, Harcourt, Brace and Company, 1945.

Hurt, K. F. : *The Quest of the Psychological Jesus*, Unitarian Universalist Christian, 1982.

Jack-Roller, The, and Snodgrass, Jon: *The Jack-Roller at Seventy, A Fifty-Year Follow-Up*, Lexington, Mass, Lexington Books, 1982.

Joel Moses: *The Computer Age, A Twenty-Year View.*Cambridge, MA, The MIT Press, 1979.

Jones, Andrew: *Building a Digital Forensic Laboratory*. Butterworth-Heinemann. 2008.

Joyce, G. C. : *The Inspiration of Prophecy,* An Essay in the Psychology of Religion, New York, Oxford University, 1910.

Keane J.: *Democracy and Cyber Crimes,* London, Verso, 1988.

Kunkel, F. : *Creation Continues, A Psychological Interpretation of the Gospel of Matthew,* New York, Paulist, 1987.

Laliberte, S. : *Security by Example Defend I.T.* New York, Addison Wesley, 2004.

Marshall, Angus M.: *Digital forensics: digital evidence in criminal investigation.* Wiley-Blackwell. 2008.

McMahon, David: *Cyber Threat: Internet Security for Home and Business.* Warwick Publishing Inc.: Toronto, 2000.

Peter T.: *Manager's Guide to Internet Security.* CSI: San Francisco, 1994.

Reich, W. : *Character Analysis,* Orgone Institute Press, New York, 1949.

Schroeer, Dietrich: *Science, Technology, and the Cyber Crimes,* Ontario: John Wiley & Sons, 1984.

Scott M.: *Managing IP Networks with Cisco Routers.* O'Reilly & Associates, Inc.: Sebastopol, 1997.

Sellin, Thorsten: *Culture Conflict and Crime,* New York, Social Science Research Council, 1938.

Siegel, Larry J. :*Criminology.* California: Thomson Wadsworth, 2005.

Smith, Susan J.: *Crime, Space and Society,* Cambridge, England, Cambridge University Press, 1986.

Susan J.: *Crime, Space and Society,* Cambridge, England, Cambridge University Press, 1986.

Tillich, P. : *The courage to be,* Yale University Press, New Haven, CT, 1932.

Travis Hirschi: *A General Theory of Crime,* Berkeley, California, Stanford University Press, 1990.

Uday O. and Vijay K. Gurbani: *Internet & TCP/IP Network Security.* McGraw-Hill: New York, 1996.

Verton, D. : *Black Ice: The invisible Threat of Cyber-terrorism,* New York: McGraw-Hill/Osborne, 2003.

William Aspray: *Computer, A History of the Information Machine,* New York, Basic Books, 1996.

Wilson, C. : *Information Operations and Cyberwar, Capabilities and related Policy Issues,* In C. R. Service, Congressional Research Service, The Library of Congress, 2006.

Wolfgang, Marvin E.: *Patterns in Criminal Homicide,* Montclair, New Jersey, Patterson Smith, 1958.

Zanini, M.: *Countering the New Terrorism,* Santa Monica, CA: RAND, 1999.

# Index

## A
Actuarial Assessment, 133.
Assessment of Trauma, 130.
Automated Fingerprint Identification System, 34, 163.

## B
Biological Approach, 80.
Biological Theories in Criminology, 80.

## C
Child Marriage, 91.
Cognitive Assessment, 126.
Cognitive Development Theory, 166.
Competency Assessment, 127, 132.
Competency Evaluations, 11.
Computational Criminology, 77.
Computer Forensics, 13, 14, 16, 17, 18, 19, 27, 29, 30.
Consequences of Crimes, 94.
Conservative Criminology, 78.
Crime Assessment Stage, 65.
Crime Scene Analysis, 99.
Crimes against Women, 90.
Criminal Behaviour, 80, 81, 82, 83, 84, 85, 87, 88, 89, 97, 98, 99, 137, 138, 139, 140, 141, 142, 143, 145, 146, 147, 148, 149, 150, 151, 152, 153, 154, 155, 158, 166, 178, 179, 182, 183, 184, 190, 191, 194, 195, 196, 200, 201, 202, 203.
Criminal Mind, 160.
Criminal Profile Stage, 68.
Criminal Profiling, 73, 99, 100, 170.
Criminal Psychology, 73, 90, 91, 198.

## D
Device Forensics, 13, 27, 28, 29, 31, 47.
Digital Forensics, 8, 13, 14, 15, 18, 20, 21, 24, 25, 26, 27, 28, 44.
Diminished Capacity, 49, 51.
Domestic Violence, 90, 91, 181.

## E
Environmental Influences, 138, 141, 142, 143, 144, 145, 149, 150.

## F
Fear of Crime, 53, 54, 55, 77.
Forensic Evidence, 5, 17, 41, 121.
Forensic Process, 13, 20, 23, 29.
Forensic Psychology, 1, 2, 3, 5, 6, 9, 46, 76, 85, 141.
Forensic Psychology Practice, 10.
Forensic Science, 4, 5, 6, 7, 8, 13, 14, 16, 17, 19, 35, 42, 43, 44, 46, 48, 49, 50, 51.

## G
General Theory of Crime, 195, 196, 202.

## I
Investigation of Crime, 184.
Investigation Stage, 68.

## J

Justifying the Punishment, 76.

## L

Learning Theories of Crime, 199.
Learning Theory, 143, 144, 150, 153, 158, 167, 168, 169, 199, 200, 203.
Liberal Criminology, 79.
Liberal Cynical Criminology, 79.
Litigation Science, 4.

## M

Methods of Punishment, 76.
Mitigating Circumstances, 47.
Mobile Device Forensics, 13, 27, 28, 29, 31, 47.

## N

Nature of Criminology, 76.
Network Forensics, 13, 18, 19, 20, 21, 28.

## O

Offender Profiling, 160.
Offender profiling, 73.
Offender Typologies, 177, 178, 180.

## P

Peer Influence, 86.
Personality Disorders, 85, 88, 115, 138, 146, 147.
Psychoanalytic Theory, 166.
Psychogenic Approach, 80.
Psychological Profiling, 173.
Psychological Testing, 101, 102, 108, 135.
Psychological Tests, 10, 102, 103, 109, 119, 122, 124, 125, 134, 135, 136.
Psychological Theories of Crime, 166.
Psychometric Testing, 109, 111, 114.
Public Perceptions, 55.

## Q

Questionable Techniques, 10.

## R

Radical Criminology, 79.

## S

Sanity Evaluations, 12.
Sentence Mitigation, 12.
Serial Murder Cases, 50.
Sexual Harassment, 91, 92.
Social Structure Theory, 157.

## T

Test of Memory Malingering, 127, 130.
Therapeutic Evaluation, 3.

## V

Validity Indicator Profile, 127, 130.
Victim Facilitation, 95, 96.
Victim Proneness, 94.
Violence Risk Assessment, 119, 127, 133.
Violence Risk Assessment Guide, 127, 133.

❏❏❏